NEW DIMENSIONS IN AGRICULTURAL GEOGRAPHY

VOLUME TWO

THE ECOLOGY OF AGRICULTURAL SYSTEM

Concept's International Series in Geography

Volume 1	:	**Historical Dimensions of Agriculture**
Volume 2	:	**The Ecology of Agricultural System**
Volume 3	:	**Socio-economic Dimensions of Agriculture**
Volume 4	:	**Landuse and Agricultural Planning**
Volume 5	:	**Spatial Dimensions of Agriculture**
Volume 6	:	**Human Perception and Technological Change in Agriculture**
Volume 7	:	**Dynamics of Agricultural Development**
Volume 8	:	**Anthropogenic Dimensions in Agriculture**

Concept's International Series in Geography No. 4

NEW DIMENSIONS IN AGRICULTURAL GEOGRAPHY

VOLUME TWO

THE ECOLOGY OF AGRICULTURAL SYSTEM

NOOR MOHAMMAD
Editor

CONCEPT PUBLISHING COMPANY, NEW DELHI 110059

ISBN 81-7022-400-4 (Set)
ISBN 81-7022-402-0 (Vol. 2)

First Published 1992

Published by
Ashok Kumar Mittal
Concept Publishing Company
A/15-16, Commercial Block, Mohan Garden
NEW DELHI 110059 (India)

Lasertypeset by
Microtech Advance Printing Systems (Pvt.) Ltd.
H-13, Bali Nagar,
NEW DELHI 110015

Printed at
Sunil Printers
CB 386, Ring Road, Naraina
NEW DELHI 110028

PREFACE

AGRICULTURE CONTINUES to be the key industry. It is not only the dominant form of employment but also forms the basis for a way of life for, perhaps, two-thirds of the world's households. Agriculture being the function of physical, socio-institutional and techno-economic factors which are dynamic in nature, keeps on changing with the basic objective of increasing production and generation of foodgrains surplus. The traditional system of farming is in the process of transformation to modern agriculture which itself is undergoing a set of equally and generally even in more rapid changes. So it has added a new dimension in agriculture which needs to be studied and analysed.

A project was, therefore, taken up to accomplish this task. Original research papers dealing with various facets of agriculture were invited from scholars of different discipline from all over the world. The response was very encouraging and a large number of original papers were received. Besides these, some papers which have classical and everlasting value have also been picked up from my earlier publication – *Perspectives in Agricultural Geography*. The papers have been classified according to their basic themes and have been put into following eight volumes so as to give a systematic picture of various themes. The articles dealing with concepts, theories and models have been kept first followed by those dealing with the same theme at macro, meso and micro levels respectively.

The First Volume – *Historical Dimensions of Agriculture* – containing 19 articles deals with the origin, diffusion and development of agriculture, research trends in agricultural geography, world agricultural systems and methods and techniques in agricultural geography.

The Second Volume – *The Ecology of Agricultural System* – deals with the selected parameters of physical environment specially the climatic elements which affect the agricultural landscape in spite of technological advancement in agriculture.

The Third Volume – *Socio-economic Dimensions of Agriculture* –

containing 19 articles deals with socio-economic environment which have bearings on agricultural systems. The first 8 articles deal with social parameters specially labour, caste, women etc., while the last 11 articles examine the impact of techno-economic factors specially the various facets of irrigation which is a basic agricultural input.

The Fourth Volume – *Landuse and Agricultural Planning* – contains 19 articles dealing with concepts, theories, models of landuse, land capability and suitability, wastelands and dynamics of landuse.

The Fifth Volume – *Spatial Dimensions of Agriculture* – includes articles on agricultural systems of specific areas such as Himalayas, Bastar, Uttara Khand etc. It also deals with dynamics of crop landuse, cropping pattern and its diversification, crop ecology and crop-combination analysis.

The Sixth Volume – *Human Perception and Technological Change in Agriculture* – contains only 15 articles and covers four basic themes. The first set of 3 articles deals with human perception and his consequential response, while the second set includes articles on correlates of technological change and adoption of agricultural innovations. The third set of articles deals with the assessment of socio-economic transformation in rural India. The last set deals with agriculture and human settlements.

The Seventh Volume – *Dynamics of Agricultural Development* – includes articles on concept of development and its various facets. Besides assessing the trends of levels of agricultural development, it also examines and evaluates its impact on environment and ecological parameters.

The Eighth Volume – *Anthropogenic Dimensions of Agriculture* – containing 17 articles deals with regional imbalances in agricultural productivity, methods of its measurement, and also its correlates and determinant. It also focuses on food availability and food security at regional and continental level.

The collection of papers in eight volumes of *New Dimensions in Agricultural Geography* is designed to encapsule the major themes and recent trends and concepts in the field of agricultural geography with particular reference to conceptual revolution in the discipline. All the volumes contain a brief introduction which gives a glimpse into the topics dealt with in each of them.

The variety of articles contributed by the renowned scholars of different disciplines has made the character of these volumes multi-disciplinary and has enhanced their value. These volumes have been designed to help not only the students of a particular discipline but also to general students who wish to keep abreast of the latest developments in the field of agricultural geography. From these points of view these volumes hold a promise of rich fare and in addition to being heuristic, will provide a mine of information for discerning scholars. It is hoped that this unique collection of research papers will not only be valuable to the researchers in the fields of agricultural geography, economics, sociology, history, and agricultural science, but will also provide guidelines to decision-makers and planners.

The completion of this project has been possible due to the constant inspiration and unstinting co-operation of a host of people too numerous to acknowledge individually. However, I owe a debt of gratitude to my mentor — Dr. Mohammad Shafi, Professor Emeritus, Department of Geography, A.M.U., Aligarh who gave me valuable comments and suggestions ever since the inception of this project. Prof. R. Ramachandran and my other colleagues who offered sincere comments on the title of volumes and the collection of articles deserve special thanks. I would also like to thank my father, wife and children who provided a favourable working atmosphere to me for successful completion of this project. Last but not the least, I would like to put a special word of appreciation, love and affection to Mohammad Imran Ali Khan who is a source of hope and inspiration.

NOOR MOHAMMAD

Department of Geography
Delhi School of Economics
University of Delhi
Delhi-110007
January 26, 1992

PUBLISHER'S NOTE

When Concept's International Series in Geography No. 1 — *Perspectives in Agricultural Geography* edited by Dr. Noor Mohammad in five volumes was issued over a decade back, we expressed our apprehensions that 'Posterity will be the best judge of our efforts in this direction. The response to this set of volumes will provide the impetus for hastening the publication of further series that are already on the anvil'.

Perspectives in Agricultural Geography received such an overwhelming response that it became out of print soon after its publication. Its success provided the necessary encouragement to issue thereafter Concept's International Series in Geography No. 2 — *Perspectives in Urban Geography* edited by Dr. C.S. Yadav in 16 volumes and No. 3 — *Perspectives in Geomorphology* by Dr. H.S. Sharma in four volumes.

We are glad to offer now Concept's International Series in Geography No. 4 — *New Dimensions in Agricultural Geography* edited by Dr. Noor Mohammad in eight volumes. In view of the new emerging trends, concept, theories and models in the field of agricultural geography, the set presents a wide panoramic view of the latest advances in this field. Hope this set too would be equally welcomed by the students, researchers, scholars and planners in various disciplines.

February 11, 1992 NAURANG RAI

CONTENTS

LIST OF FIGURES

LIST OF PLATES

LIST OF TABLES

LIST OF CONTRIBUTORS

Abbunoori, Abbas Ali

Formerly Research Scholar, S.V. University, Tirupati

Ahmad, Enayat

Professor of Geography (Retd.), Department of Geography, Ranchi University, Ranchi

Basu, Subhashranjan

Department of Demography and Applied Geography, University of North Bengal, Siliguri

Bhagabati, Abani Kumar

Department of Geography, Gauhati University, Guwahati

Das, M. M.

Department of Geography, Gauhati University, Guwahati

De, N. K.

Department of Geography, Burdwan University, Burdwan

Gupta, R. K.

Central Soil and Water Conservatiion Research and Training Institute Dehradun

Hayter, R.

Department of Geography, Simon Fraser University, Burnaby, Canada

Kar, Amal

Central Arid Zone Research Institute, Jodhpur

Maunder, W.J.

Superindent of Special Services, New Zealand Meteorological Services, New Zealand

McDonald, G. T.

Division of Australian Environmental Studies, Griffith University, Nathen, Brisbane, Queensland, Australia

Mohammad, Noor

Department of Geography, Delhi School of Economics, University of Delhi, Delhi

Mooley, D. A.

Indian Institute of Tropical Meteorology, Ramadurg House, Pune

Raza, Moonis

Professor Emeritus, Centre for Study of Regional Development Jawaharlal Nehru University, New Delhi

Singh, Arun Kumar

Human Ecologist, Anthropological Survey of India, 2, Rippon Street, Calcutta

Singh, Harjit

Centre for the Study of Regional Development, Jawaharlal Nehru University, New Delhi

INTRODUCTION

THE STUDY of influence of environment upon the nature and distribution of man's activities and qualities is the specific field of geography (Visher, 1932). Nature with all its diversities in its physical characteristics — relief, climate, soil, drainage and natural vegetation, provides a host of different possibilities to man to adopt in different parts of the earth surface, and man with his existing level of technological advancement has availed these opportunities (Hetner, 1947). However, of all the economic activities, agriculture is, probably, one on which the phenomenal environment impinges most (Hurst, 1972). Hence, from the very beginning the cultivators were concerned with the problems explaining as to how the variation in the environment affected the agricultural landscape (Gregor, 1970). In fact agriculture is not merely growing of crops but it is more than that viz., it is also a form of applied ecology. Crop landuse and to a lesser degree animal husbandry, both depend on immediate natural environment which can be changed only at heavy cost particularly in arid and semi-arid regions of the world (Mithorpe, 1965). McHarg (1966) on the basis of principles of ecological determinism has shown how nature can impose limits, provide guidelines and assist man in solving environmental planning problems in a manner which is most compatible with natural problems. Hence, efforts need to be made to quantify the environmental attributes to meet various human needs by establishing basic guidelines for future ecological planning, modification and adjustments. Brookfield (1969) has drawn attention to many problems in measuring the perceived environment because it is complex, monistic, distorted and discontinuous, unstable and full of irrelevancies.

However, during the past three decades man, due to his technological advancement, has overcome many of the problems imposed by nature on agriculture. Consequently, tremendous development has taken place in agriculture which has resulted in ecological disturbances leading to a severe threat to sustainable human prosperity. This phenomenon has recently been realised and various studies have been conducted to assess the impact of agricultural development on environment. **McDonald** in his paper has suggested system approach for assessing the impact identification, perdiction and evaluation. The author has also suggested that before implementing any agricultural development project a detailed study of its positive and negative impact must be done under a specific guideline and if it is absolutely positive, the project should be inducted.

A proper understanding of climatic environment in relation to crop, and thereafter identification of agro-climatic zones, is essential for suitable cropping pattern and improvement in agricultural production specially in a rainfed area. **Abbunoori** has made this attempt in the drought prone areas of Iran where the drought conditions influence agriculture in a much stronger way than anywhere else. The author has identified agro-ecological zones on the basis of index of moisture status and water balance and has tried to associate the crop production and moisture index. For an indepth study wheat and barley have been taken up and the fluctuation in their area, yield and production is worked out for the period 1970-71 and 1983-84 at the interval of one year for each of the agro-ecological zones. Finally, a brief account of index of moisture adequacy and cropping pattern is given. This article is a maiden attempt, on agriculture of the arid regions of Iran and its findings may be valid to other such regions of the Middle East.

Bhagabati and **Das** in their article have assessed the performance of agriculture in three distinct ecological zones of Nalbari district viz. (a) migrant-dominated active flood-plain, (b) indigenous non-tribal dominated built-up middle plain, and (c) indigenous tribal dominated high plains. The study is based on primary data collected from six villages selected on the basis of stratified purposive sampling. The authors have assessed the variation in landuse pattern in terms of the ecologico-social space, land-holding size and scattering of agricultural plots, area and yield of major crops cultivated in different spaces, and the pattern of adoption of modern inputs and implements in agriculture. The study reveals that the pattern of agricultural performance varies significantly with the variation in the ecologico-social spaces. The immigrant-dominated flood-plain is at a higher level in total performance of agriculture, while the indigenous non-tribal dominated plain is at the lowest.

Human response towards nature is generally region specific depending on resource base, the level and the nature of the natural constraints and that of technology available. **Singh** in his paper examines this problem at micro level in Ladakh region where nature expresses its hostility through rugged mountainous terrain and cold arid climate. The region has been divided into two parts on the basis of human response to inclement environment: the pastoral area lying around Changthang above 4500 metres above mean sea level, and agricultural area lying in river valleys below 4500 metres. The author has analysed the extent and nature of constraints on agricultural economy in the second region. He has critically studied with the support of empirical data, the impact of relief — altitude, slope and aspect, climate-temperature, rainfall, and other associated phenomena. The farming calendar shows that the number of growing days is very limited. However, the study confirms that inclement climate and rugged mountainous terrain are the major constraints in the agricultural development of Ladakh region which may be true to other such regions of the world. The author has suggested other measures to develop the economy of this region.

Although natural phenomena are far from explaining everything, yet slope, soil, climate, hydrographic conditions etc. have very important effect and are responsible in general for the orderly, often much too orderly, of human phenomena. Generally the man's genesis with his rare and skilful flexibility adapts itself to the most varied conditions, but the abnormal and unexpected events, in fact, do overwhelm and paralyse him. **Basu's** article is tentative and speculative inquiry at macro level into cultural causation, centering its attention on the issue of physical processes in origination and change in Bhagirathi Valley. The study is based on field investigation and published data. A substantive portion of the paper is devoted to the analysis of geomorphological evolution and characteristics of the valley specially the fluvial characteristics. The process of terrace formation which reflects the topographic variations has been given intensive treatment. The topographic forms of ups and downs in causation to their textural variations in soil have undergone three essentially different cultural landuse. The upper terraces of loamy sand are utilised for homestead surrounded by orchards. The lower terrace of silty loam are best suited for market gardening. Interculture is also favoured. The lowest approach of silty clay is prepared for paddy cultivation.

The concept of land for agriculture is not limited within the boundaries of a particular field of a cultivator. It takes into account the total geomorphic study of the agricultural field including the slope of the land, morphological characteristics and sequence of the land drainage characteristics. Hence, any planning for agriculture should assess first of all the physical potential of the land and how different geomorphic parameters react under any induced condition. Overlooking of these factors may sometimes prove to be disastrous for agriculture (Michel, 1972). This problem has been studied by **Kar** by taking examples of Nile valley, Sindh plain, Punjab plain, Ganga delta, Assam region and Amazon basin where unscientific planning has proved disastrous in the long run. In spite of all these glaring examples, still wrong decisions are taken and faulty programmes are formulated for the development of agriculture. The development of agriculture does not necessarily mean full scale change in the existing pattern nor does it necessarily involves high sounding and costly changes in the morphological features of the land. What is needed is better understanding of the land forming processes, the physical potential of the land, and how the different land form units react to the external processes. Based on this, new projects should be planned in such a way as not to adversely affect the ecological balance, but to reap our harvest keeping the harmonious relations with the land.

Geomorphology is one of the dominant parameters of physical environment and its impact on patterns and intensity of agriculture is so immense that it could be seen at macro, meso and micro levels. Agriculture in mountains, plateaus and plains is different from scarps, pediments, valley flanks, flood-plains, ridges, alluvial flanks, drumlines, monodnocks, leaves, dunes and oases. **Ahmed**'s article is a conceptual one and it deals with impact of geomorphology on agriculture. The author has discussed the role of macro and micro regional geomorphology on agriculture and has highlighted their impact on soil moisture, depth, structure and texture of soil, amount of solar radiation, the feasibility of terracing, field pattern, agricultural transports, possibilities of irrigation, size and distribution of holdings, continuity of cultivation, crops, their nature, yield and outturn. He stresses on slope and terracing, and discusses as to how the degree of slope affects the farming. Besides, he has discussed thoroughly and systematically the influence of geomorphology on field pattern, ground-water and agricultural water supply, nature and contribution of crops and agricultural transport with suitable examples taken from all over the world.

Deserts have fragile ecosystem where varying degree of aridity and terrain constraints put severe limitation to agricultural and allied activities. Terrain constraints are viewed as land degradation process and associated natural hazards which are related to morphodynamic process operating over the various landforms. **Kar** in his paper has accounted these processes during Quarternary period and Recent time together with the human activities accelerating these processes creating fluvial erosional, aeolian, salinity and alkalinity hazards in the Indian Desert. The author has recommended the need for understanding the process-form interaction, classification of hazards for mapping, utility of ramote sensing in mapping the present status and monitoring the changes.

Pastoral farming is highly susceptible to the variation in weather conditions and it results in fluctuation of pastoral production from season to season. And, herein lies the significance of weather forecast. **Maunder** in his paper has discussed the advantages of weather forecast and its implications to the transport industry and other activities beyond the farm gate in New Zealand. Since the whole agricultural chain from the farm to the market place is considered as an integral system, it is imperative to establish a methodology for accurately establishing the volume of pastoral products so that export earners function at their peak efficiency through the whole pasture to market chain. It requires the close liaison between the metro-system and pastoral producing industry in order to fully realise the significance of weather variations to pastoral industry. The author has examined various weather based production models regarding national dairy production and wool production. The study reveals that weather conditions both actual and forecast, can be of considerable value in forecasting pastoral production in New Zealand. However, researches on economic value of weather is only useful as it affects the way decisions are made by managers of weather sensitive processes. A continuing need exists, therefore, to assess and present the impact of weather in terms of production figures, costs or similar measures, which can be used directly by decision-makers including economists, agronomists, planners, transportation corporations and politicians. If it is done, the farming community and the nation will reap the benefits.

Climate, to a very large extent, still determines where man may live and thrive, what type of house he may appropriately built, what sorts of clothing he needs and what pests and diseases he must combat (Whitebeck, 1932). Although man's influences — from stand point of historical, political and economic factors, have been great, climatic factors still are basic in determining the distribution and performance of crops. With evergrowing mastery over environment, the climatic elements, though can be modified but only at some cost, play a dominant role (Coppock, 1971). The capability of producing a crop in a region is dependent mainly on existing climatic and soil conditions. As the climatic factors mainly exert their influence on plant life, the differences in the behaviour of crops may be considered primarily due to difference in climate rather than soil condition (Klages, 1958). Similar views are put forth by many scholars (Stamp, 1967; Thomas, 1967 and Dumont, 1970).

However, **Mooley** has given a systematic and analytical view of the Indian summer monsoon, its impact on agriculture and remedial measures. The author has reviewed the genesis of the Indian monsoon system. The influence of the Indian mountain system on the monsoon, contribution of the monsoon to the annual rainfall, monsoon seasonal rainfall, and economic benefits of the monsoon have been discussed. Abnormal behaviour of the monsoon during the period 1871 to 1978 has been examined. The consequences of the abnormalities and the remedial measures which can be adopted, have been considered. An examination of the monsoon rainfall for the period 1871 to 1978 has revealed that though the Indian monsoon rainfall may depart substantially from normal and lead to suffering by the people on a large scale in a few years, it has a tendency to return to the normal in a couple of years and that there is no long-term adverse trend to cause any worry. The occurrence of the years of large scale monsoon failure and of large scale monsoon rainfall excess is found to be random and the number of years of such monsoon failure or of such monsoon rainfall excess in a five-year period follows Poisson probability law.

The influence of climate is most conspicuous in traditional peasant agriculture of underdeveloped regions. The characteristics of the soil are largely the product of present and past climates and the vegetation that flourishes in them, and the effect of relief are to no small degree expressed through resulting climatic variations (Symons, 1978). The paper of **Das** deals with this problem at meso level in the state of Assam. The climate of Assam is characterised by modified tropical monsoon with well-marked seasonality. There is heavy rainfall in summer and continuous drought in winter. These two seasons are separated by pre-monsoon and autumn seasons which are transitional in characteristics between the former two seasons. About 70 per cent of the total rainfall of the year occurs in the months of July and August.

The agricultural activities of the peasants are conditions by the seasonal rhythm of the climatic conditions. But there is spatial variation in the distribution of climatic elements within the state. Accordingly cropping pattern also varies from region to region. On the basis of climatic variations the state can be divided into six climatic zones which are exactly identical with the cropping pattern of the different regions. The paper also attempts to establish a *prima facie* case in favour of the crucial need of artificial irrigation for successful agriculture in different parts of the state.

Frost is an other weather element which affects the agriculture most. The protoplasm in living plants can function properly only within a restricted temperature range. No growth of agricultural importance takes place when temperature approximates the freezing point (Chang, 1971). However, in Canada frost is both a constant limitation to the range of agricultural practices, and a killing hazard which can occur sporadically during the growing season. The ultimate occurrence of frost results directly in substantial losses of crops in occasional years as well as in deterioration of crops quality. The breeding of varieties of cereals which can mature in shorter frost free periods is major part of the battle to push cultivation further north in northern Europe, Siberia and Canada (Symons, 1978). **Hayter** and **Proudfoot** have discussed the problem of frost hazard and its effect on agriculture in the continuously settled part of the province of northeast of Edmonton. The study area is characterised by undulating terrain which is deçply incised in restricted area by the Athalia sea and north Sasketchwan river. The response of individual farmer to the frost hazard is the function of not only of their awareness and knowledge of the incidence of frost and its effects but also of their assessment of other physical factors which govern profitability. The authors have computed mean and median values of several stations to ascertain frost free season. They have probed further to study the micro-climatic variations at several representative stations. In the study area, with its varied topography, the principal micro-climatic variations are associated with cold air drainage, resulting in temperature variations, the presence of large water bodies which may completely counteract the mechanism of cold air drainage and vegetation. The study reveals that within agricultural area northeast of Edmont there is no general regional variation in frost occurrence and in the length of the frost free period. Instead, micro-climatic variations are significant.

Weather and climate influence the productivity and marketing economy of primary sector to a large extent. It is more so in agriculture specially in case of dairy farming and livestock ranching which produce perishable materials. Weather forecasting is, therefore, very significant in operational decision-making. Hence, it is the responsibility of the meteorologists to provide most appropriate weather information (past, present and future), to the users. **Maunder** in his paper has discussed the use and significance to weather information in making operational decisions and has considered it as a challenge to meteorologists. He has critically evaluated the significance of weather information for planning and overall production in a region. The author has elaborated as to how weather based predictions of dairy production assist the New Zealand Dairy Board in assessing more accurate market and shipping requirements. Likewise he has also evaluated its use and significance in electric power and planning, sheep and lamb production and planning. In fact the basic factor in optimising the use of meteorological information is their appropriate weighting in order to make them available in the form appropriate to users. This has been done according to the distribution of thirty-four economic and agricultural parametres. They are currently analysed for 24 areas of New Zealand and these informations are available in real time.

Among all the parameters of physical environment, climate, perhaps, plays a determining role in shaping the agriculture and its productivity in a region. Hence, for a rational agricultural planning for higher agricultural production there is a need for detail information of various climatic elements. **Mohammad** and **Raza** in their paper have assessed the impact of availability of meteorological data on agricultural productivity. The location of Agricultural Meteorological Station (hereafter AMS) is taken as representative of availability of agricultural meteorological data in India. The study is based on secondary data obtained from Indian Meteorological Department for AMS and Raza's (1981) article for agricultural productivity. Since the AMS differ from one another in variety of ways, the authors have given scientific scores to various kinds of AMS in order to make their availability in different regions comparable. For the assessment of spatial distribution of AMS the country has been divided into Land Potential Regions based on **Gupta** *et al*, paper (1981) and the rank of each region is ascertained according to the AMS scores. It shows that semi-medium potential regions rank high followed by very high and very low potential regions respectively which indicate that there is no relationship between the land potential regions and distribution of AMS. For ascertaining the relationship between AMS and agricultural productivity the country has been divided into four agricultural productivity regions and the rank of each region is ascertained with respect to AMS which shows very high degree of position relationship between the two except in case of very low productivity region which ranks third in place of fourth. The association becomes even stronger when we take the areal strength of each productivity region.

Mountains are generally inaccessible and are not suited for better human habitation. Hence, are sparsely populated homes of selected isolated tribes who utilise mountain ecosystem for subsistence activities on a marginal level. But there are certain tribes who utilise the mountain ecosystem/resources in a more scientific way and Apa Tanis of Lower Subansiri district of Arunachal Pradesh is one of them who may be considered as model rice cultivators in mountain ecosystem. **Singh** in his paper has examined in detail the nature and characteristics of Apa Tanis society in the environmental framework of Lower Subansiri district, the evolution of rice cultivation among them, types of rice fields, methods of rice cultivation, and water management practices. The tribe has, in fact, adopted a very scientific system of rice cultivation which is intuned with the overall ecological system of the area. Hence, it supports a very high density of population due to high agricultural yield in a small area without causing any serious damage to environment. It has also been made possible due to self-containness of the society both politically and ideologically and also due to sustained peace.

Floods and waterlogging create acute problems for the development of agriculture. Considerable property is damaged due to devastating floods in rural as well as in urban areas. Hence, top priority is given in planning to control floods in agricultural regions specially in flood prone areas. But there are certain regions where flood menace is unavoidable. In such areas flood mitigation projects are introduced. **McDonald** in his paper has developed a model to assess such projects on the basis of empirical data derived from the report that assessed the hydrological, economic and environmental impacts of a project on the Macleay river flood plain in New South Wales (Australia). In fact the problem of planning an appropriate combination of structural and non-structural measures to mitigate flood damages in rural areas in Australia have received a great deal of considerations but difficulty still exists in providing suitable technical data to allow adequate evaluation of alternatives and to indicate the extent of hydrological change in the planning regions. Agricultural geographers can help in this work. Of course, other more intricate problems exist such as scaling the long-term economic significance of any benefits, accounting for secondary effects and forecasting the impact of changes on the long term equilibrium of flood plain habitats. However, the author in his project assessment model has included thirteen components and for each component he has recommended research techniques. Finally, he has suggested that for assessment application four types of analysis could be undertaken viz. (i) assessing the distribution and extent of flooding characteristics, (ii) comparing the changes sustained before the installation of mitigation work with that existing now, (iii) measuring the changes brought about by parametric variations in the design capabilities by mitigation work, and (iv) commenting on possible changes in other environmental variables and their effects on the other functions.

Soil is the source of practically all man's food, clothing and an ever increasing list of other needs. Actually man gets nearly all his food from soil, less than one per cent of what he eats being fish (Pearson and Harper, 1945). Hence, geographical investigation of soil characteristics in agricultural geography is of great significance. There are different aspects of soil and its uses but **De** has taken only fertility and productivity in his paper. The concept of fertility and productivity is very complex and is used in different sense in different context. But in brief fertility is the quality that enables the soil to provide the proper compounds in proper amounts and in the proper balance for the growth of the specific plants, when other factors such as light, moisture and physical conditions of the soil are favourable; while soil productivity is the resultant of various factors influencing soil management. Thus soil productivity is a function of environmental factors combined with soil fertility. However, the author has carefully examined in detail the factors affecting soil fertility such as soil depth, soil texture, soil pore-space, soil temperature, soil aeretion, soil compactness and tillage, soil reaction, and nutrient contents. The factors controlling soil productivity have also been discussed and grouped into two categories viz. inherent productivity factors and different soil condition influencing productivity.

Generally land capability, which refers to as an expression of all the environmental parametres apply to biological potential of a given resource from a specific unit of land and which takes into consideration primarily the soil parametres such as top soil, effective soil depth, associated land features e.g. slope of land, extent of erosion, degree of wetness and susceptibility to overflow and flooding, is taken as basis for regional landuse planning. But **Gupta** in his article advocates that due to peculiar socio-economic constraints such as size of holdings, dependence of large population on agriculture, an ecological approach to land capability classification should be adopted and used for landuse planning. The author after an elaborate explanation of the concept with suitable examples has critically reviewed in detail the land capability classification done in some typical areas of the country such as Sutlej Beas catchment area and Naurar catchment area in Kumaon hills, and has suggested more scientific bases for land capability classification and conversantly landuse planning.

REFERENCES

Brookfield, H. C., 1969 : 'On Environment as Perceived' in *Progress in Geography*, Vol. I Board, C. et al. (eds).

Chang, J. H., 1971 : *Climate and Agriculture: An Ecological Survey*, p. 100.

Coppock, J. T., 1971 : *Agricultural Geography of Great Britain*.

Dumont, R., 1970 : *Types of Soil Economy: Studies in World Agriculture*, Vol. 3.

Gregor, H.F., 1970 : *Geography of Agriculture: Themes in Researches*, p. 31.

Gupta, et al., 1984 : 'Land Resource Regions and Areas of India' in A. Ramesh (ed), *Resource Geography*, Heritage Pub., Delhi.

Hetner, A., 1947 : *Allgemien Geographic Des Menchen* (General Geography of Humanity), Vol. I, p. 1.

Hurst, M. E. E., 1972 : *A Geography of Economic Behaviour : An Introduction*, p. 83.

Klages, K. H. W., 1958 : *Ecological Crop Geography*, p. 44.

McHarg, I. L., 1966 : 'Ecological Determinism' in M.F. Darling, and J. Milton (eds), *Future Environment of North America*, p. 190.

Michel, M., 1972 : *The Impact of Modern Irrigation Technology in the Indus and Helmund Basins of Southwest Asia*.

Mithorpe, F. L., 1965 : 'Crop Response in Relation to the Forecasting of Yield' in Johnson, C. G. and Smith, L. P. (eds), *The Biological Significance of Climatic Change in Britain*, pp. 119-128.

Pearson, F. A. and Harper, F. A., 1945 : *The World's Hunger*, p. 5.

Raza, M. 1981 : 'Regional Disparities in India : A Preliminary Exploration of the Regional Dimensions of Agricultural Development' in N. Mohammad (ed), *Perspectives in Agricultural Geography*, Vol. IV, Concept Publishing Co., New Delhi.

Stamp, L. D., 1967 : *Applied Geography*, p. 15.

Symons, L., 1978 : *Agricultural Geography*.

Thomas, M. F., 1967 : *Farming in Hot Countries*, p. 15.

Visher, I. S., 1930 : 'Recent Trends in Geography', *Scientific Monthly*, Vol. 35, p. 439.

1

ENVIRONMENTAL ASSESSMENT FOR SUSTAINABLE RURAL DEVELOPMENT

G. T. McDONALD

OVER THE past two decades the world has experienced agricultural progress but at the price of substantial increase in the use of fertilisers and pesticides, greater dependence on engineered plant and animal varieties, more intensive use of land by multiple cropping and a shift towards monoculture. Food production has increased but the number of people who are hungry has increased and the pockets of severe inequity remain. The natural resource capital in terms of undeveloped lands, forests and water supplies is declining.

The possibility exists now that development itself is a threat to sustainable human prosperity. Certainly there is growing concern to assess more carefully new development projects, the unforeseen consequences of pesticides, forest exploitation, soil erosion and land degradation, desertification, salinisation of irrigation projects and environmental health problems resulting from rural development.

Environmental impact assessment (EIA) is an essential component in planning for *sustainable production* from rural areas. Pressure on rural lands for economic growth coupled with the complexity and damage potential of modern biological and chemical technology presents a difficult challenge to environmental assessment.

There is also a need to lessen, and hopefully reverse, environmental decline in traditional farming areas where land degradation, reduced productivity, over-used water supplies and the increasing magnitude of effects from natural hazards such as drought and flood are apparent.

Apart from meeting agricultural objectives in terms of food security and regional development for rural populations, rural lands must also serve the needs of forestry, mining, rural settlement and the increasing demands of modern cities such as water supply and infrastructure, energy developments and to assimilate greater quantities of wastes. Maintaining gene pools and ecological processes such as watersheds and harvestable wildlife habitat are important functions of the rapidly declining stock of natural lands in rural areas.

EIA grew out of concern about the unintended consequences of development projects and technological change. Assessing specific projects and technologies has contributed a great deal to the minimisation of negative effects and maximisation of the beneficial effects of development.

Of all the rural development projects, land development, clearing natural vegetation for agriculture has the most far-reaching environmental consequences. New settlements are created, new industries established and an existing natural system replaced by a managed one. A great deal has been written on the environmental assessment of the many land development projects throughout the world (Lal, Sanchez and Cummings, 1986).

There are many methodological difficulties involved in carrying out impact identification, prediction and evaluation. One useful approach to EIA is to adopt a systems approach and this is described briefly below in Table 1.1:

TABLE 1.1: Two decades of world agricultural development

Region	Per Capita Food Production (1964=100)		Per Capita Gross Cropped Area (ha)		Per Hectare Fertiliser (kg)	
	1964	1984	1964	1984	1964	1984
World	100	112	0.44	0.31	29.3	85.3
N. America	100	121	1.05	0.90	47.3	93.2
W. Europe	100	131	0.31	0.25	124.4	224.3
E. Europe, USSR	100	128	0.84	0.71	30.4	122.1
Africa	100	88	0.74	0.35	6.9	53.6
W. Asia	100	107	0.53	0.35	1.8	9.7
E. Asia	100	116	0.30	0.20	6.4	45.8
S. America	100	108	0.49	0.45	11.6	32.4
CPEs of Asia	100	135	0.17	0.10	15.8	170.3

Source: *Our Common Future*, Report of the World Commission on Environment and Development.

Agricultural Systems Analysis : A Basis for Impact Identification, Prediction and Evaluation

The Value of a Systems Approach

In recent years agricultural scientists and economists have successfully applied the concepts of systems analysis to problems related to the development of the agricultural sector in many countries. A system is a set of components connected by flows of energy, matter, control etc. For the natural components such as soils, plants and animals, this is referred to as an agro-ecosystem; for the farm management aspects, a farming system. In reality these are parts of a larger system which is called an agricultural system.

While the concepts of systems analysis are potentially very complex and the requirements for analysis very demanding, as *a way of thinking*, agricultural systems analysis has a great deal to offer environmental impact assessment. It emphasises the number of different components in an agricultural system, of the connections between these components, of system dynamics over space and time, explicit boundaries, of critical flows of a bio-geochemical nature, of the linkages in food chains and of the role of the farmer in these processes. In the socio-economic arena, it focuses our attention on the inter-relatedness of agricultural practices, the farmer, farm families and villages and broader marketing and social issues.

An example of an agro ecosystem that illustrates these points is the rice-field (Fig. 1.1). The water-retaining dyke or bund forms a strong and easily recognisable boundary, while the irrigation inlets and outlets represent some of the limits outside linkages. The great diversity of wildlife in the original natural ecosystem is reduced to a restricted assemblage of crops, pests and weeds. The basic ecological processes such as the competition between the rice and weeds, consumption of the rice by pests and the predation of the pests by their natural enemies remain, but are now overlain by the agricultural processes of cultivation, artificial inputs, control and harvesting.

The same kind of picture can be drawn for higher levels of agricultural systems, for the farm, for the village or the watershed, but the increasing complexity of the interactions makes a simple graphical representation difficult.

System thinking helps with the identification of many of the important short term impacts of a project on local flora, pests, fertility, soil erosion, labour requirements, market effects etc. It is particularly useful in progressing beyond the identification of direct effects to indirect effects.

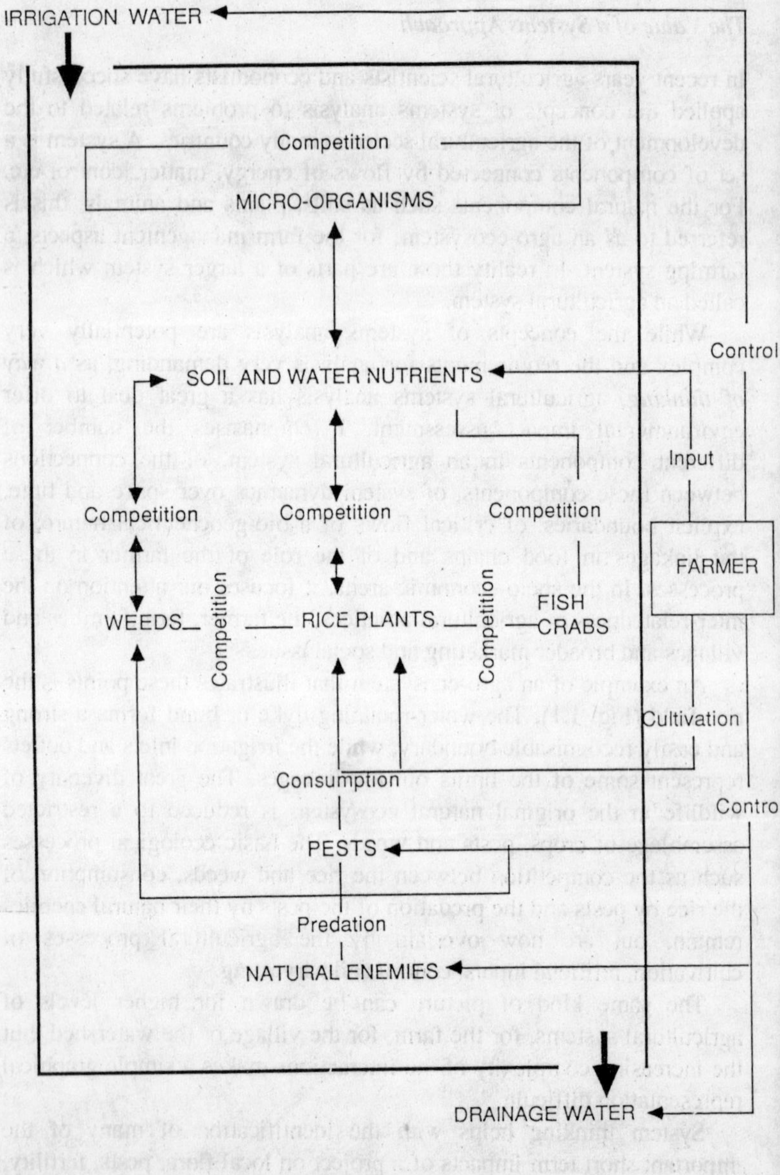

FIG. 1.1: The ricefield as an agro-ecosystem.

For scientists, analysis of the components of an agricultural system may lead to quantitative *prediction models*. Of course this may only be justified for very large projects, although scientific progress provides generalisations about the operation of hydrological systems, soil erosion, bio-geochemical processes, markets etc. that are of value even when there is relatively little data available.

Human activities affect biota more drastically by affecting processes rather than through the extermination of populations as such. Furthermore, these *processes* change less from place to place than does the distribution of species.

Evaluating System Performance : Key Properties

Systems thinking draws our attention to the interrelationships between components and their dynamic properties. Four system properties together, describe the essential behaviour of agro-ecosystems (Fig. 1.2). They are :

(1) *Productivity*: the net increment in output per unit of input, commonly measured as annual yield or net income.

(2) *Stability*: the degree to which productivity remains constant in spite of normal small fluctuations in controlling environmental variables, such as climate, or in economic conditions, normally measured as relative variability.

(3) *Sustainability*: the ability of a system to maintain its productivity when subjected to stress or perturbation. A stress is a regular, possibly continuous, small and predictable disturbance, for example the effect of increasing soil salinity or farm indebtedness. A perturbation is by contrast, an irregular, infrequent, relatively large and unpredictable disturbance such as is caused by a drought, a new pest or a major political upheaval. Unfortunately, measurement and prediction of perturbations is difficult and can often only be done retrospectively. Lack of sustainability in a system (collapse) may be sudden, without warning, or it may come gradually.

(4) *Equitability*: how evenly the products of the agricultural system are distributed amongst its human beneficiaries. The more equitable the system the more evenly the agricultural products, the food or the income are shared amongst the farm village or nation.

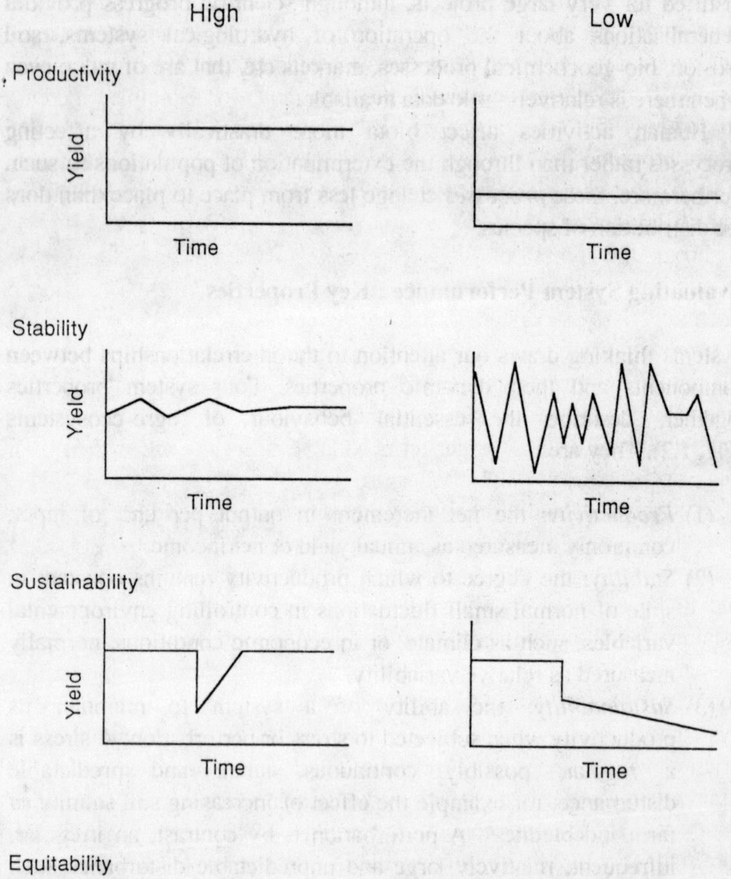

FIG. 1.2: The properties of agro-ecosystems.

These indicators of the system are descriptive, but they can also be used to compare systems, or in the context of impact assessment to evaluate the desirability of alternative courses of action. In agricultural development projects there are almost always changes in these fundamental properties and the need for a trade off between them. New forms of agriculture or new techniques may, for example, increase productivity, but often at the expense of lowered values for one or the other properties. There are almost inevitably trade-offs between productivity and stability on the one hand and sustainability and equitability on the other.

From the point of view of impact assessment, changes in productivity will almost always be regarded as desirable, indeed essential. Risks associated with stability may be regarded as acceptable in a modern state, but sustainability, a much longer term issue, is a dilemma. To what extent should a project be developed now which cannot be guaranteed to survive; one that is known to have potential flaws that may lead to longer term collapse at a time when there is a greater population at risk?

Traditional agricultural agro-ecosystems such as swidden generally have low productivity and stability but high equitability and sustainability (System A in Table 1.2). Traditional cropping systems tend to be more productive and stable yet retain a high degree of sustainability. Equitability varies. The introduction of new technology (for example new high yielding rice varieties) while greatly increasing productivity is likely to lead to lower values on the other properties (System C). Yields fluctuated widely for early high yielding rice varieties which were more prone to pest and disease attack. The requirements for cultivation favoured only some farms, often those that already had significant economic advantage. More recent varieties are more stable but have poor sustainability.

Identifying Agricultural Systems

The task of identifying the key components in an agro-ecosystem is basically multi-disciplinary and should also involve local extension workers and farm leaders. After defining the main components through a workshop with these people, several organising principles can be used:

– location — what is happening where — transects are

particularly helpful in this.

- time — What is happening when — crop calendars are very useful in this.
- flows — what are the main flows of energy, water, materials (including wastes) and of money.

In an impact assessment project, the accuracy of the identification and prediction of impacts will depend on how much time and resources are available for research. The research questions will be forthcoming from the identified flows and components. If rapid rural appraisal is all that can be afforded, then the systems description, as a focus for workshop discussion will produce a most cost effective impact assessment.

TABLE 1.2: **Agricultural development and agro-ecosystem properties**

System	Productivity	Stability	Sustainability	Equitability
A. Swidden Cultivation	low	low	high	high
B. Traditional Cultivation	medium	medium	high	medium
C. Improved Scientific Methods	high	low	low	low
D. Ideal Methods on the Best Land	high	medium	high	high

Source: Conway (1985).

Socio-economic Impact Analysis : Farming System Research

Similar issues arise with focus on the human dimension of an agricultural system. Farming system research is farm management research with people dimension added.

In order to identify the components of a farming system it is possible to use "rapid rural appraisal" where the emphasis is upon understanding the main components of an agricultural system and the processes rather than to attempt for unattainable quantitative accuracy in measuring the existing farming structure. Rapid rural appraisal is, as the name suggests, a "methodology" used for low cost data acquisition in rural development planning. It has considerable application to impact assessment. There is no prescriptions for this methodology which might include some, or all of the following :

- air photograph and remote sensing data (for area wide landuse classification and enumeration)
- interviews with community leaders and local officials (for issues and social processes)
- intensive surveys of a very small number of representative farms (to establish basic elements of farming systems, factors, inputs and agricultural processes)
- transects (to establish/confirm landuse patterns and practices and their topographic relationships)
- published statistics.

Unfortunately there is no recipe book for the conduct of a systems analysis, only that art-form of applied science that comes from experience of knowing what is relevant.

The "adaptive environmental assessment and management" model of Holling is based on system thinking and the creation of multi-disciplinary teams to analyse processes.

Project Specific Assessment

The Asian Development Bank (ADB) Guidelines provide a useful model for the process of environmental assessment of rural projects. The basic purpose of the Guidelines, which have been prepared for many classes of projects, is to assist in the preparation of the Initial Environmental Evaluation (IEE) for a proposed project prior to the initial feasibility study. This is really a screening procedure. If the IEE indicates that a fuller Environmental Impact Assessment (EIA) is required, the IEE is of considerable assistance in determining the Terms of Reference for the EIA (scoping) and in suggesting appropriate mitigation strategies (Fig. 1.3).

An Impact Checklist is used to identify potentially significant impacts, both positive and negative, and to grade them according to their significance. If no significant impacts are identified, then no follow-up action is required. Where significant impacts are identified

Action Affecting Environmental Resources and Values	Hazards to the environment if not done
ELEMENTS OF SUCCESSFUL ECONOMIC-ENVIRONMENTAL LAND CLEARANCE	
Reliable estimate of the agricultural productivity of the lands to be cleared	Hazard to the economic viability of the project
Consistency with national forestry policy	Long term loss in national economy and welfare
Appropriate sizing and location of area to be cleared	To minimise disturbance to natural forest areas
Selection of optimal clearing method	Serious hazard of negation of project benefits and efficiency
Optimal post-clearing operations	Serious hazard of negation of project benefits and efficiency
Economic and social evaluations	Essential to ensure project viability
Appropriate use of agro-forestry	Loss of chances to enhance overall project benefits and appropriateness
Reclamation of swamplands/careful evaluation of alternatives of filling versus aquaculture	Loss of chances to enhance overall project benefits and appropriateness
Roads — careful planning and construction	Hazards of silt run-off and encroachment into forests
Use of degraded or unused lands	
MONITORING	
Construction (clearing stage)	Construction contractor likely not to observe constraints
Post-clearing stage	Operating personnel not likely to manage project at optimal efficiency

FIG. 1.3: Checklist of environmental parameters projects in land clearing and rehabilitation.

1. This lists the significant environmental effects known to occur in land development projects in developing countries.
2. This is arranged to permit:
   (i) Ready screening out of non-pertinent items.
   (ii) Ready grading of significant effects by degree of effect.
3. The checking process (2) furnishes the information needed for preparing an IEE.

Source: Environmental Guidelines for Development Projects: Land Clearing. Asian Development Bank Manila. May 1987.

which can be mitigated, further recommendations of study may be required. Where significant impacts have been identified an EIA is required on those impacts.

This procedure, apart from having obvious administrative benefits to the ADB and other agencies, provides :

- the *checklist* as a convenient and efficient starting point for the impact assessor, focussing attention on key issues that have been collected from a range of similar projects in the region. This is a saving in research time for the busy assessor.
- *criteria* to assist grading the degree of likely impacts. While these are not extensive, again they provide guidance gained from experience to assist the assessor.
- examples of potential *mitigation measures* to encourage good environmental practices from the very beginning of the project cycle. Often mitigation measures can reduce negative impacts significantly without affecting project costs or timing (e.g. minor shifts in location of processing plant, dry season land preparation, adequate drainage of roads etc.).

The general principle of environmental guidelines for development planning is very sound. The guidelines benefit specific administrative arrangements and given the positive tone (What is really important? What can be done in advance to avoid problems?), the communications with other professionals and administrators is enhanced. Some bad projects will never see the light of day because the proponents have some clear idea as to the environmental sensitivities of their projects.

The logical extension of this kind of process is to have "class assessments" for many types of relatively minor projects. These assessments could be self-administered by agencies responsible for the particular class of project (e.g. roads, small irrigation works, solid waste dumps, water harvesting etc.). They are in some countries.

For land development and other similar projects it is my belief that, wherever possible, cost/benefit analyses (CBA) should be used. Despite the great difficulties of applying CBA to EIA, it is essential to use the logic of CBA to prevent misleading costs and benefits being claimed. One trivial example is that all benefits should be in net terms (returns - costs). Claiming gross output as a benefit is a common way distortion.

Landuse and Environmental Planning

Difficulties in resolving competing demands for rural land and resources and interdependencies between environmental assessment and regional planning logically lead to the incorporation of project assessments within integrated resource use plans.

The point is often reached in preparing a project assessment where the assessment of impact is contingent upon :

- future commitments of land and other resources in the project area
- acceptable environmental losses of natural lands, of increased populations and of the consequences of degradation within a framework of national and regional resource use and conservation objectives.

The EIA practitioner is commonly faced with a "chicken and egg" question. A project can only be assessed within a given (or projected) context of population, economic activity and resources use and yet the project may itself be a key element in those activities through its many secondary effects. Plans cannot be prepared without knowing what projects will be built — projects cannot be assessed without knowing what other developments will take place.

The paper by Hanson and Koesoebiono (1979) on developing coastal swamplands in Sumatra, is an excellent case study on the need for integrated resource management. It illustrates very well that while impact assessment is an essential element in the development planning process, it is totally imbedded within the whole decision-making process in the area. There is a set of interdependent decisions that must be made. Assessing the impacts of, for example drainage canals, cannot be made without consideration of a whole range of other decisions about settlement location and size, farming systems, transportation developments and conservation of fisheries habitats etc.

EIA and Landuse Planning

Landuse planning is deciding in advance what to do, how to do it and who is to do it in relation to :

- the commitment of the major landuses — allocation

– the management of uses.

Along with pollution control, landuse planning is a key component of environmental planning.

Landuse planning is carried out in a hierarchy of levels with a greater attention to detail at lower levels including :

– *national landuse plans* which will deal with national economic, social and environmental policy and specific objectives including food security, sectoral allocation of capital, housing, national conservation strategies etc.
– *regional landuse plans* which will include the settlement system, land tenure and land reform, forestry, major infrastructure, such as roads and water storages, resource developments, such as land development and mining, wildlife and conservation area management
– *local landuse plans* which deal with detailed allocation of uses of land and the operational management of villages, landuse change and subdivision, water supply.

The Role of EIA

Project definition is a critical step during which the need for the plan, its goals, and the constraints on the study are defined. The way in which higher level plans are to effect planning activities, the accepted commitments of land prior to planning, the adoption of an environmental dimension are all important elements of this.

Environmental assessment enters this phase of the planning through its role in problem definition (what kinds of environmental problems exist now and are likely to exist in the future) clarifying environmental goals and standards (what are realistic expectations for the region). Unless these issues are included in the definition phase, the plan may be dominated by other considerations such as commercialisation, resource exploitation, infrastructure objectives. UNEP and other environmentally sensitive guidelines may be useful in this phase. Even ensuring that the definition phase does not lose sight to the key issues of sustainability, conservation of species and productive ecosystems is a valuable step.

The environmental evaluation component is a key to any landuse plan. An assessment of land resources together with demands on those resources is at the heart of the planning process : the search for *alternatives*. The environmental assessment dimension comes into this phase in many ways but particularly :

- an assessment of environmental resources should always be included. This will cover : (i) the identification of nationally and regionally significant conservation areas (in terms of species and ecosystems), (ii) natural systems that support economic activity such as fisheries, water supply, tourism, and (iii) environmental quality issues including water quality and waste management. Environmental values need to be assessed explicitly — what are the intrinsic conservation values of natural areas (are they rare etc.) and what are the ecological requirements of processes such as waste assimilation, water purification, fish habitat etc. Sometimes these things are omitted. All the use values of an area need to be assessed not just the value for rice, cotton etc.
- a statement of acceptable environmental impact should be incorporated in the formula for any landuse suitability. Methodologies for assessing landuse suitability always consider the requirements for production of crops, forests etc. They should be scrutinised also to ensure that suitability takes account of land degradation and the off-site effects of the use (e.g. protection of streams and watersheds and stream banks in the case of land development schemes).

There are many approaches to environmental survey for landuse planning. Once a landuse plan has been prepared, EIA is incorporated by:

- preventing uses in particular areas where their environmental impact is too high
- providing criteria for EIA on uses that are not prohibited
- indicating areas where there is no requirement for EIA for particular uses.

Agricultural geographers should apply their skills of field research and problem solving to the pressing needs of sustainable development. Whilst the EIA technique is not the perfect solution to the problem of sustainability, it is a practical approach that is widely adopted throughout the world. Agricultural geographers could make a significant contribution to sustainable development by improving our knowledge of agro-ecosystems and assessing the impacts of rural development using EIA.

REFERENCES

Conway, G. R., 1985 : "Agricultural ecology and farming systems research", in J. V. Remenyi (ed): *Agricultural Systems Research for Developing Countries*, Canberra, Australian Centre for International Agricultural Research, Proceedings No. 11.

Dobbin, J., 1981 : *National and Sub-National Plans for Sustainable Development*, Gland, Switzerland, IUCN Commission on Environmental Planning.

Hanson, A. J. and Koesoebiono, 1979 :"Settling coastal swamplands in Sumatra: A case study for integrated resource management", in C. MacAndrews and Chia Lin Sien (eds), *Developing Economies and the Environment*, Singapore, McGraw-Hill.

Lal, R., Sanchez, P. A. and Cummings, R. W., (eds), 1986 : *Land Clearing and Development in the Tropics*, Rotterdam, A. A. Balkema.

throughout the world. Agricultural geographers could make a significant contribution to sustainable development by improving our knowledge of agro-ecosystems and assessing the impacts of rural development using EIA.

REFERENCES

Conway, G. R., 1985. "Agricultural ecology and farming systems research". In J. V. Remenyi (ed.): *Agricultural Systems Research for Developing Countries*. Canberra: Australian Centre for International Agricultural Research. Proceedings No. 11.

[illegible], J. 1991. *National and Sub-National Plans for Sustainable Development*. Gland, Switzerland: IUCN Commission on Environmental Planning.

Hanson, A. J. and Koesoebiono, 1979. "Settling coastal swamplands in Sumatra: A case study for integrated resource management". In C. MacAndrews and Chia Lin Sien (eds): *Developing Economies and the Environment*. Singapore: McGraw-Hill.

Lal, R., Sanchez, P. A. and Cummings, R. W., (eds), 1986. *Land Clearing and Development in the Tropics*. Rotterdam: A. A. Balkema.

2

AGRO-ECOLOGICAL ZONES AND AGRICULTURAL PRODUCTION IN DROUGHT-PRONE AREAS OF IRAN

ABBAS ALI ABBUNOORI

KNOWLEDGE OF climatic environment in relation to crop will enable not only the evolving of a suitable cropping pattern but also the modification of various agronomic practices for better productivity and output. Identification of agro-ecological zones is a pre-requisite for the development of crop production, particularly, rainfed crop production. Needless to say that production of a particular crop in any region is the product of area under the crop and its yield per unit area. Area and yield can each be expressed as a function of a number of controlled variables, i.e., fertilisers, pesticides, irrigation, price, etc., and uncontrolled set of variables comprising various climatic factors like rainfall, temperature, humidity, etc.

A study of the interaction between the climate and agricultural production or crop-weather relationship is actually a complex problem because of numerous measurable and non-measurable parameters.

Importance of Agronomic Study

The effect of weather needs to be drawn out of the observed behaviour of crop-yields before the contribution of technology and inputs can be properly evaluated. There are no studies on agronomic theory to identify the critical phases of plant growth under the influence of different weather variables at each phase of growth.

Since it is difficult to establish crop-weather relationship for specified crops and regions, it follows that it is even more difficult to establish the relationship between output aggregated over several crops and spatial units on the one hand, and the aggregate indices of weather, on the other. In addition, no distinction is made between yield or production under irrigated and rainfed conditions. Any theory that fails to distinguish yields of irrigated from rainfed land is obviously untenable.

Influence of Drought on Crops

General shortage of water and prevalance of drought conditions are by far the most serious constraints for stabilisation or promotion of plant production in the arid and semi-arid or dryland regions of Iran. The problem is extremely complex as the plants in the open system of the field are subjected to wide ranges of soil-water conditions. However, the soil component and the plant component of the overall soil-water-plant-environment complex can be gainfully managed to some degree to improve the agricultural production. Lahiri (1980) has pointed out that drought may be of different types and may be broadly classified into four different categories. In one case, erratic distribution of rains leads to sporadic drought conditions where plants may be subjected to soil-water deficit at any stage of growth. In another situation, rains may completely fail and thus very low soil moisture may prevail throughout the growing period of the crop. In the third situation, early rains followed by prolonged dry periods may progressively dwindle the soil-water reserve and thus plant-growth and yield may be severely restricted. In the fourth situation, soil-moisture stress, particularly at the critical reproductive phase of the plant, may reduce the yield of the plant. It is not only the volume of rainfall but also its distribution at different stages of the plant growth that influences the total output. The adequacy and timeliness of rainfall at the pre-sowing period affect the sown area. Similarly, the volume and distribution of rain at the time of sowing, flowering, maturing, the harvesting affect crop-yield. Consideration of these issues would necessitate introduction of a large number of explanatory variables into the crop-rainfall relationship and thereby, reduce the degree of freedom for estimating the parameters with adequate levels of confidence. In addition, crop calendar is not the same for sowing and harvesting in different provinces and in different periods. It is influenced by the adequacy and timeliness of rainfall during the growth of the crop.

Crop failure is an extreme and transparent consequence of

inter-acting agro-climatic, biological, and soil events. Farmers exercise more control over the choice of the cropping system and the crop operations. They manipulate cropping combinations over space and time and these choices provide flexibility in dryland farm management Singh and Walker (1982). Larson and Eastin (1971), have pointed out that reduction in productivity is due to a combination of climatic factors. Water deficiencies not only reduce the yield, but also change the pattern of growth. In Iran, the rainfed cultivations are subject to periodic, partial or complete crop-failure, because of the irregular or erratic distribution of water-supply and rainfall even within the rainy season.

Agro-Ecological Zones of Iran

An attempt is made in the present study to identify agro-ecological zones of Iran on the basis of index of moisture status and water balance elements instead of mere rainfall, as it would be more rational and appropriate. This is because rainfall never fully reflects the moisture status of a region, and also plants do not fully depend on rainfall alone. Besides, the significant role played by soil as the repository of water cannot be overlooked. However, the annual rainfall in Iran, particularly in the semi-arid region, is abnormally distributed over the years and there is no clear evidence of any non-randomness in the distribution. Here an attempt is made to examine graphically the relationship between the fluctuations in crop production and productivity and the selected environmental factors, i.e., the Index of Moisture Adequacy (IMA), Departure of Aridity Index from the Median (D), and precipitation. For the sake of better analysis of the crop-weather relationship, the diagrams for climatic events are drawn separately. It is notable that the analysis of failure of rainfall as against the fluctuations in crop-production show some instances where despite the annual rainfall being below normal for quite a few provinces the crop-production has not shown any "trough". This may be due to the mutual offsetting effect in production fluctuations between the provinces, particularly in the semi-arid region.

Weather and Wheat and Barley Crops in Iran

We have tried to examine the crop-weather relationship for two major crops viz., wheat and barley. In the semi-arid region these crops are mainly grown under rainfed conditions. Both the meteorological and the crop data for each crop for each year, are represented pictorially in crop-weather Fig. 2.1. Such diagrams summarise a wealth of data depicting the history of the crop. The study period chosen for this analysis is the period from 1970-71 to 1983-84. A trend variable is supposed to include improvements in inputs, farming practices, skills, etc., the impact of which is to be expected in any particular year, and serve as the reference point which is assumed to be constant here. It can be seen from the given diagrams that there is a relationship between the output fluctuation and climatic variations specially in the semi-arid region. The year-wise levels of area, production, and yield per hectare of wheat and barley for the semi-arid provinces of Iran are presented here.

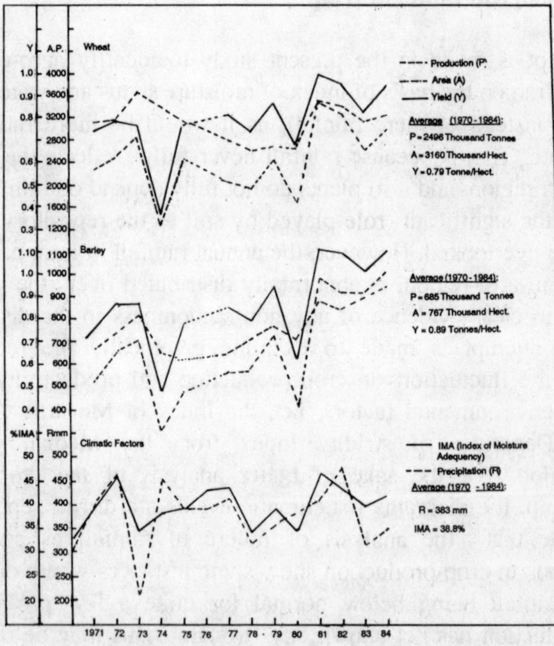

FIG. 2.1: Crop - weather relationship — Semi-arid region (Total).

In general, 1972–74 can be treated as a bad climatic period for most parts of Iran, particularly for the semi-arid provinces. The following year, 1974-75 registered a steep decline in the levels of output of wheat and barley. The intensity and frequency of climatic trends accompany crop fluctuation. However, in 1974-75 the yield of these two principal crops, i.e., wheat and barley, was considerably affected. The reduction in wheat was 53.05 per cent and barley, 45.1 per cent. Thus, from these percentages it is clear that the effect of drought on wheat is more severe than on barley. At the other extreme, 1981–83 was recognised as the best period for agricultural production. In this period the yield and production levels of wheat and barley appear to be at the highest level. In 1981-82 the yields of wheat and barley accounted for 26.9 per cent and 28 per cent above the average level. The regression analysis for this period shows a positive relationship between crop production and related characteristics of weather.

Region-wise Weather-Crop Experience in Iran

It is already noted that drought is often a factor in yield reduction even when the damage is not apparent. The climate and rainfall pattern, onset, intensity, frequency, duration, amount, cessation etc., during the crop-growth period determine the crop yield, particularly in the rainfed areas of Iran. Sometimes the crop yields are affected considerably by a short spell of very unfavourable weather if it occurs at the critical phase of crop-growth. For instance, in the Azerbaijan Sharghi province the moderate drought of 1973-74 reduced the yield of both wheat and barley in 1974-75, whereas in 1982-83, inspite of large intensity of drought and the lower amount of rainfall, the yield of wheat and barley appeared not effected severely. It is needless to say that a short moisture stress or dry spell at flowering and grain formation stage or the "critical period" of the crop may greatly reduce economic yield without any appreciable reduction in the amount of rainfall.

It should be noted that in our graphical analysis here, the departure of aridity index from the Median (D) is plotted inversely to give a clear picture of drought intensity. It should be noted that the high yield of wheat in 1984-85 was due to shrinkage of the area under this crop. In the case of barley, the lowest yield of 0.10 tonnes per hectare, i.e., 76.2 per cent less than the average, was recorded in 1974-75, and the highest yield of 0.74 tonnes per hectare, i.e., 76.2 per cent higher than the average, in 1982-83. In 1982-83, the annual rainfall recorded 444 mm which is 35.8 per cent higher than the average. In other words, the output and the climate diagrams of this province show that they are moving almost together but in different intensity.

Similarly, in the other semi-arid provinces, i.e., Kurdestan, Lorestan and Hamedan, the yield as well as output of the two major crops. viz., wheat and barley appear to be at the lowest level in 1973-74. In this year, the reduction in the yield of wheat and barley, respectively, recorded 62.7 per cent and 60.6 per cent in Kurdestan, 61.3 per cent and 38.7 per cent in Lorestan and 66.2 per cent and 57.5 per cent in Hamedan. Moreover, in these provinces the sharp decline in the yield and consequently in the output level of these two crops follow the climatic trend and show a deep downward movement. At the other extreme as in the other semi-arid provinces, the highest yield of wheat and barley was recorded in the year 1981-82. Thus, in these homo-climatic provinces of the semi-arid category, not only there is similarity in crop fluctuation but also a close relationship between crop fluctuation and climatic trends. This study is to bring out the fact that in the semi-arid region the production and productivity of wheat and barley crops are a function of environmental and climatic factors. An attempt is made here to consider a few agronomic factors that are helpful in alleviating the effect of drought on a few crops with special reference to wheat and barley.

The importance of water stress during the early stages of reproductive plant growth was examined by Salter and Goode (1967), who examined 132 reports on the effect of water deficit on the yield of cereals. Over one hundred of these reports show that yield was affected most adversely when water stress occurred during the periods of internode elongation just before flowering, ear-emergence from the leaf sheath, and flower-opening. In most crops, the adverse effects of drought vary in intensity according to the stage of growth. It is, therefore, clear that the critical stage of development of the crop when it is most susceptible to drought, should not coincide with the period when drought usually occurs. In this regard Chinoy (1960) says that the greater susceptibility of late flowering varieties to drought during their shooting and flowering stages is not due to any hereditary differences in their drought-resistance capacity.

The effect of water stress on plant growth has been discussed in several reviews (Kramer, 1949; Slayter, 1957, 1960; Kozlowski, 1958, 1962, 1964; Kramer and Kozlowiski, 1960; Vaadia et al, 1961; Slavik, 1965; Kozlowiski, and Keller, 1966; etc.).

TABLE 2.1: The critical periods of spring cereals in relation to their growth stages and phases (After Skazkin, 1961)

Wheat

Growth stage	Seedling	3rd leaf	tillering	shooting	ear formation	Flowering
Growth phase	Vernalisation Phase		light phase	3rd phase	4th phase	
Time of critical period					Critical period	

Barley

Growth stage	Seedling	3rd leaf	tillering	shooting	ear formation	milk	Wax
Growth phase	Vernalisation Phase		light phase	3rd phase	4th phase	ripeness	ripeness
Time of Critical period					Critical period		

Under yield consideration, certain experimental approaches have been used to determine the effect of either a plentiful supply of water or of drought conditions at different stages of subsequent growth and yield of the crops and using statistical methods of study correlations between rainfall over particular periods and crop yields. Tinsley and Vernon (1904 and 1905) believed that irrigating more frequently than once every three weeks after heading had begun increased yields of wheat but the increase was scarcely enough to cover the cost of extra irrigation. In drought or water-stress situations, improved yields can be obtained by irrigating a crop during the critical response period. Corresponding growth stages and phases have been outlined in relation to critical response periods by Skazkin (1964) (Table 2.1).

Wheat will not germinate below 4°C and new seed may not germinate above 15°C, although there are cultivars available which will germinate upto 29°C. Cool, moist, conditions are ideal for both crops but they will tolerate high temperatures when established, provided the humidity remains low. Barley is more adaptable to high temperatures than wheat and is better suited for the arid sub-tropical coastal plains in the south of the country where it is more likely to be used as a livestock fodder rather than a cereal grain. Both crops require a minimum growing period of 95 days but the average growing period for a cereal crop in Iran is 220-250 days with the maximum period in the coolest areas exceeding 300 days.

Among the other scholars who have carried out a considerable number of experiments to determine the effect of reduced soil moisture supply or proper time for irrigation of these crops are: Moliboga (1927), Welch (1914, 1917), Kezer and Robertson (1927), Robertson et al (1934), Palmer (1939), Skazkin et al (1935), Nesterova (1935), Slabonski and Lipinska (1960), Hutcheon and Rennie (1960), Hooker (1907), Fisher (1924), Hallgren (1947).

In the case of Barley, Musaeva and Aspinall studied its critical period and Salter concluded that the experimental results indicate that there is a moisture-sensitive period in the growth of barley when deficiency in soil moisture results in reduction in grain yield which cannot be offset by optimum moisture conditions at other periods. This moisture-sensitive period occurs at the end of the shooting stage and during the earing when the generative organs are being formed. During this period drought conditions have their maximum effect in reducing yield, and irrigation has the greatest effect in increasing yield, and excess water has the smallest adverse effect on yield.

In general the simplest, most common, and generally accepted classification is:

Drought Resistance – Drought Avoidance + Drought Tolerance. Levitt et al. (1960) and Levitt (1972) have greatly amplified and subdivided this simple scheme. Before considering mechanisms for avoiding or tolerating drought, we should consider 'drought escape' which is the most effective adaptive mechanism concerning productivity. 'Drought escape', selected in all rainfed cereal systems, aims to completely preclude drought from interacting with crop growth and yield. The result of such selection is a cropping season in which all yield determining growth stages are carried out during the periods of very low probability of drought Oka, (1958), Vergara and Chang (1976), Krishnamurthy, et al. (1971).

Summary of 'Critical Response' Periods Covering All Cereals

The following conclusions can be drawn regarding the response of the cereal crops of soil moisture conditions at different stages of development :

(i) These crops show a marked sensitivity to water supply during the formation of the reproductive organs and during flowering,

(ii) Shortage of water at this time can seriously reduce grain yield through a reduction in the number of grains formed per year,

(iii) The tissues of the reproductive organs appear to be particularly susceptible to damage from water-shortage, and pollen-formation and fertilisation can be seriously affected by such conditions,

(iv) Root growth of many of these crops is very much reduced and may even cease when the flowers start to develop. Such a reduction in root activity coming at this critical time can result in a serious reduction in water absorption by the plant unless the soil moisture content is sufficiently high to permit rapid water-movement to the root surface, and

(v) This critical stage of development very often coincides with almost total depletion of the supply of available water in the soil and with a maximum transpiration demand. Thus the chances of the plant being subjected to water shortage at this time are high, especially if atmospheric humidity is also low at this time.

Irrigation has a maximum beneficial effect on yields when it is given during the development of the floral organs. Rain has a similar beneficial effect at these stages of growth.

In all cereals water deficit at the reproductive stage is potentially the most damaging to the yield (Salter and Goode, 1967, Fisher, 1973, Boyer and Mc Pherson, 1975 and Begg and Tunner, 1976).

Index of Moisture Adequacy and Cropping Pattern

The Index of Moisture Adequacy, defined as the ratio of actual evapotranspiration (AE) and potential evapotranspiration (PE), is a useful gauge of plant-water supply in relation to plant-water demand. Water is essential to plant growth and most of the needed water is stored in the soil. The availability of this water depends on the water balance of the soil which in turn depends on precipitation (replenishing) and evapotranspiration (exhausting).

The elements of water-balance obtained from the book-keeping procedure when plotted graphically against the months of the year, present a clear picture of the marches of not only the water surplus and water deficiency through the year but also reveal the accumulation and depletion of soil moisture. The 40 per cent isoline (annual) of the index seems to delimit zones favourable for crop culture, unless otherwise raised as irrigated crops. Data below illustrates the IMA values and the corresponding crops that can be successfully grown under such IMA conditions in the absence of any supplemental irrigation :

IMA %	Crops
80 - 100	Paddy (High Yields)
60 - 80	Paddy (Low Yields)
below - 60	Paddy (uneconomical)
40 - 60	Millets
20 - 40	Droughts resistant crops
below - 20	Unsuitable for cultivation

The concept of the above Index of Moisture Adequacy (IMA) is considered here for the purpose of classifying agricultural drought and evolving a cropping pattern. For evaluating the Index of Moisture Adequacy (IMA), the water-balance is computed for about 30 years using 43 selected stations of Iran, on an annual and seasonal basis. High values of the index are found in the Gilan and Mazandaran provinces. These are the areas of fairly heavy and prolonged precipitation and consequently they rarely experience moisture deficiencies at least on an annual basis. Since large values of Moisture Adequacy Index are due to the availability of abundant quantity of moisture in relation to water need, in such regions intense agricultural operations through the year can either be expected or be recommended. On the other hand, in regions of low indices, agricultural activity is normally restricted only to the rainy season.

Fairly high values (between 40% to 80%) are found only over North West of Iran and in the rest of the regions the values are less than 40 per cent everywhere. On a seasonal basis, during March–May, the values are low everywhere (less than 40%) except in the North West where values of index are between 60 and 80 per cent. During the Winter season (Dec.–Feb.), the values of index are everywhere greater than 80 per cent. On the other hand, during the summer season (June–August) except in Gilan province the values of index are less than 40 per cent throughout the country.

From the annual AE/PE map of Iran it is obvious that rice cultivation can be successfully carried out in the North (Gilan and Mazandaran provinces), bounded by isoline of 60 and 80 per cent. In the other parts of Iran supplemental irrigation would be absolutely essential if rice has to be grown (Fig. 2.2). In this context, it is needless to say that, while rice has high thermal and moisture needs, wheat requires low temperatures and has lower water needs and is, therefore, grown in areas bound by 40 and 60 per cent isolines or lower values of index.

Thus, this Index is very useful in determining suitable crops that may be grown with the least hazard of damage from droughts. The seasonal variation of the index can be used for selecting proper locations and favourable periods for different crops.

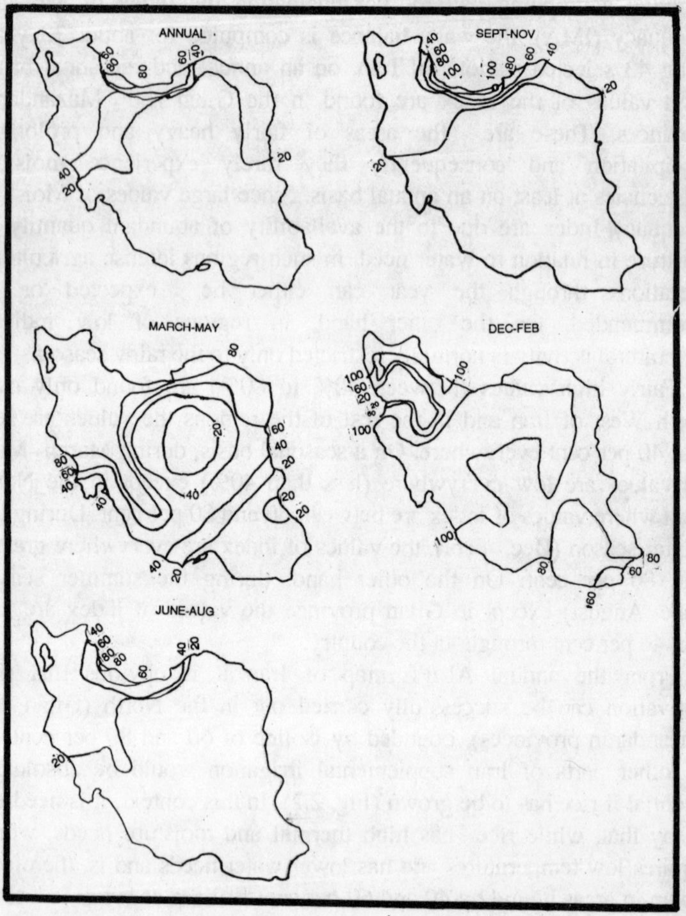

FIG. 2.2: Index of moisture adequacy.

REFERENCES

Aspinall, D., The effects of soil moisture stress on the growth of barley, II Grain Growth. *Aust. J. agric. Res.* 16, pp. 265-75.

Chinoy, J. J., 1960 : Physiology of drought resistance in wheat, I. Effect of wilting at different stages of growth on survival values of eight varieties of wheat belonging to seven species, *Phyton* 14, pp. 147-57.

Lahiri, A. N., 1980 : 'Soil-water-plant relations and production optimisation from arid areas', in H. S. Mann (ed.), *Arid Zone Research and Development*, International Symposium on Arid Zone Research and Development. CAZRI, Jodhpur, India, Printed by Harward Press.

Larson, K. L. and Eastin, J. D. (ed.), 1971 : *Drought Injury and Resistance in Crops*, CSSA Special Publication No. 2. Crop Science Society of America, Madison, Wisconsin.

Levitt, J., 1972 : *Responses of Plants to Environmental Stresses*, New York, Academic, pp. 697

Levitt, J., Sullivan, C. Y. and Krull, E., 1960 : *Bull. Res. Counc. of Israel*, 8D, pp. 173-180.

Mohan, H. S. Ram, and Subrahmanyam, V. P., 1983 : 'Application of Water Balance Concepts in Agricultural Planning and Operations in South India', *The National Geographical Jor. of India*, Vol. XXXIX, Parts 1 and 2, March-June, pp. 30-34.

Musaeva, L. D., The effect of moisture deficiency at various stages of barley growth on respiration (Russian), *Fiziol. Rast*, 4. pp. 234-42, (from *Soils and Fert.*, 21, No. 537).

Salter, P. J. and Goode, J. E., 1967 : *Crop responses to water at different stages of growth*, Res. Rev. Commonw, Agric, Bur., No.2, Farnham Royal, Bucks, England.

Singh, R. P., and Walker, T. S., 1982 : *Determinants and Implications of Crop Failure in the Semi-Arid Tropics of India*, Economics Program Progress Report-40. ICRISAT, Hyderabad, India.

Skazkin, F. D., 1964 : The critical period in plants as regards insufficient water supply (Russian), Timirjazavskie Ctenija Akab. *Nauk SSSR*, 21 p. 51 and in English Translation *Nat. lend Lib.* RTS 2355, p. 48.

Tinsley, J. D. and Vernon, J. J., *Soil moisture investigation for the season of 1903 and 1904*. Bull. N. Mex. Agric, Exp. Stat. 1904 (48) and 1905 (58).

[illegible] values [illegible] 142-15.

[illegible] Research and Development, International [illegible] Research and Development [illegible] Taiwan [illegible]

[illegible] 1922 [illegible] New York [illegible]

[illegible]

[illegible] Applied [illegible]

[illegible]

[illegible] Institute [illegible]

[illegible]

3

AGRICULTURAL PERFORMANCE IN DIFFERENT ECOLOGICAL ZONES

ABANI KUMAR BHAGABATI AND M.M. DAS

THE NALBARI District manifests itself as one of the agriculturally most problem-ridden districts of Assam. With an area of 2,102 km^2 and a population of 9.36 lakh in 1981 (projected), the district accounts for 2.67 per cent of the state's total area, but contains 4.70 per cent of the total population. Urban population constitutes only 2.2 per cent of the total population of the district and is distributed only in two small towns. Of the total workers, as high as 76.58 per cent are engaged in primary activities and the shares of secondary and tertiary sectors are 6.09 and 17.33 per cent respectively.

For a length of 87 kms, the district stretches from the riverine tract of the Brahmaputra to the Lesser Himalayan foothills on the north and cuts across a diverse physical and cultural landscape. The three broad east-west physiographic zones, viz., the active floodplain of the Brahmaputra, the middle built-up plain and the *Bhabar-Tarai* belt, each dominantly inhabited by a distinct social group, have explicit impact on the agricultural practice in the district.

The active floodplain including the *Chars* (loose sandy riverine areas) with occasional marshes is predominantly occupied by the Muslim immigrants coming from erstwhile East Pakistan (present Bangladesh). On the other hand, the middle plain having a relatively dense road network and a railway with Nalbari and Tihu towns as the central foci exhibits a very intensely humanised landscape. Here the indigenous non-tribals form the major section of population. Beyond the middle plain towards north, i.e., in the *Bhabar-Tarai* zone (zone of alluvial fans, interstream highlands and marshy plains), the landscape is marked by increasing presence of tribal population, their economy and culture. The Bodo group of tribals (Mongoloid) is concentrated in this zone. However, the habitats of the Bodos, in the recent years, have greatly been distributed by the increasing intrusion of land hungry non-tribal people from the high density middle plain and the flood-affected active floodplains of the south. Thus on the basis of the ecological and social characteristics, the district can broadly be divided into three ecologico-social spaces : (1) immigrant-dominated active floodplain, (2) indigenous non-tribal-dominated built-up middle plain, and (3) indigenous tribal-dominated high plain.

In the background of the above ecological and social characteristics, the present study attempts to analyse the performance of agriculture encompassing a number of aspects like landuse, land-holding structure, area and yield of crops and the use of modern inputs and implements in agriculture. Thus the main objectives of the study are :

1. to analyse the variation in the landuse pattern in terms of the ecologico-social spaces;
2. to study the land-holding size and scattering of agricultural plots;
3. to analyse the area and yield levels of major crops cultivated in different spaces;
4. to assess the patterns of adoptation of modern inputs and implements in agriculture.

Methodology

With the above objectives in view, a field survey with the help of a schedule was carried out during 1985-86 in six villages of the Nalbari District (Fig. 3.1). The location in respect of ecological set-up and the distance from urban and market centres, the socio-economic make-up in respect of community-wise population composition and economic condition, and the infrastructural facilities constitute the bases of selection of these villages. The villages are selected in such a way that they can be considered as the representatives of the three major ecologico-social spaces identified within the district on the basis of the ecological and social characteristics (Table 3.1).

Selection of Household

From each of the representative village, 20 per cent of the households are chosen for investigation. The households are selected with the objective of getting a representative picture of the overall agricultural situation of the villages. The procedure of household selection and investigation is as follows :

1. collection of land-holding data for all the households of the village from the *mandal* (village level field surveyor) and the *gaonburha* (village head);
2. classification of the households according to land-holding sizes (in hectare): 1, 1-2, 2-4, and > 4;
3. selection of 20 per cent households at random proportionally from each of the size-classes;
4. filling in the schedules through direct queries and also indirect assessment wherever necessary.

TABLE 3.1: Distribution of the representative villages in the major ecologico-social spaces of the Nalbari District

Village	Location in ecological zone	Dominant social group	Ecologico-social space
Kaplabori (No.2)	Active floodplain	Immigrant Muslim	Immigrant-dominated floodplain
Amayapur Kaithalkuchi Borajol	Middle built-up plain	Indigenous non-tribal (Borajol-mixed population)	Indigenous non-tribal dominated built-up plain
Dalbari Kachubari	High plain (with *Bhabar-Tarai* character)	Indigenous tribal	Indigenous tribal dominated high plain

Source: Prepared by the Authors.

The data so obtained are summarised and presented in tables for their meaningful discussion.

Introductory Profile of the Villages

As has been presented in Table 3.1, the purposively selected villages, viz., Kaplabori (No.2), Amayapur, Kaithalkuchi, Borajol, Dalbari and Kachubari represent the major ecological set-up and associated social groups of the Nalbari District (Fig. 3.1). Located on the bank of the Brahmaputra, Kaplabori represents the general character of the floodplain ecology of the Brahmaputra river. It is predominately inhabited by the socio-economically backward Muslim immigrants.

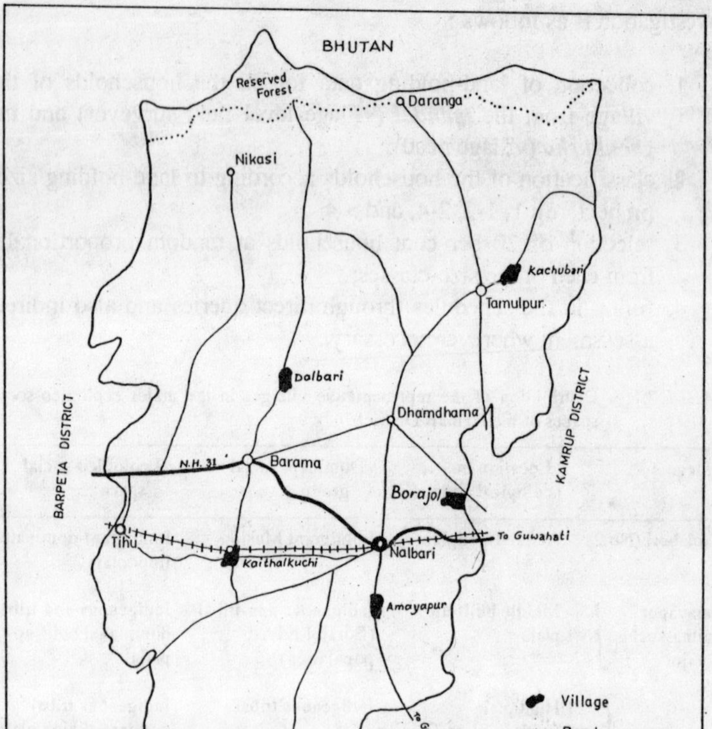

FIG. 3.1: Location of the representative villages: Nalbari district.

Amayapur, Kaithalkuchi and Borajol, being located at varying distances from urban areas and inhabited mostly by indigenous non-tribal people, represent the characteristics of the middle built-up plain. In these villages, a considerable proportion of workers is engaged in tertiary activities. The literacy level is also at a relatively high level. On the other hand, dominated by tribal economy and culture, Dalbari and Kachubari are characterised by lower proportion of tertiary workers, low level of literacy and thus, reflect the general character of the northern high plain area of the district. Table 3.2 gives the salient introductory information of the representative villages.

TABLE 3.2: Introductory information of the representative villages

Village	Area in hectare	Population	Workers in occupation (in %)			Literacy (in %)
			Primary	Secondary	Tertiary	
Kaplabori	304	2,100	83.00	3.00	14.00	22.00
Amayapur	149	1,350	46.51	3.94	49.48	70.00
Kaithalkuchi	513	3,980	59.21	6.72	34.07	47.00
Borajol	416	2,751	72.22	2.66	25.12	39.53
Dalbari	215	1,353	96.00	1.50	3.50	26.00
Kachubari	275	1,167	81.00	4.37	14.63	27.75

Source: Field work, 1985-86.

Landuse

For the analysis of landuse pattern, data are collected from the household on the heads — *basti* (settlement) land, cultivated land, fallow land and cultivable wasteland. Thus the analysis of landuse is confined only to the land held directly by the households. Table 3.3 indicates the variation in the distribution of land under different uses.

TABLE 3.3: Area under selected landuse categories as percentage of total individually owned land

Village	*Basti*	Cultivated land	Fallow land	Cultivable waste
Kaplabori	7.56	72.39	0	20.05
Amayapur	22.83	73.41	2.65	1.11
Kaithalkuchi	18.65	76.51	3.39	1.51
Borajol	17.77	74.96	6.52	0.75
Dalbari	11.69	85.65	0	2.6
Kachubari	15.16	80.20	2.72	1.92

Source: Field work, 1985-86.

The *basti* land constitutes the lowest proportion (7.56%) in Kaplabori followed by Dalbari (11.69%). A major part of Kaplabori which lies inside the new embankment of the Brahmaputra is regularly inundated by floods. The inhabitants, therefore, have selected the slightly raised lands for building their houses. Unlike the indigenous non-tribals, the Muslim immigrants who are traditionally accustomed to reverine environment are not much interested in raising spacious *bastis*.

In the middle plain villages, there exist the typical Assamese *bastis* with almost all possible physical resources needed by a rural family. In fact, the inhabitants of these villages treat their *bastis* as productive economic units. The *bastis*, particularly the larger ones, possess within them a small tank *(pukhuri)*, a kitchen garden *(saktoli)* and a bamboo grove *(bahani)* in addition to a spacious residential tract. Thus the *bastis* are found to occupy a relatively high proportion of land (17.77 to 22.83%) in the middle plain villages.

In the northern high plain villages, the tribals have also started treating the *bastis* as important sources of income and making them spacious in the recent years. Now in their *bastis* also a variety of economic plants such as bamboo, banana, areca-nut, betal leaf, coconut, etc. are generally grown. However, in both the high plain tribal villages — Dalbari and Kachubari, the *basti* land still constitutes low proportions (11.69 and 15.16%) compared to the middle plain villages.

The cultivated areas assume great significance in all the villages as their economy is primarily based on agriculture. The proportion of cultivated area does not vary significantly from village to village. Kaplabori has the lowest percentage (72.39%) of cultivated area. It is mainly because of the existence of a large amount of waste land in the village measuring 20.60 per cent of the total land owned by the households investigated. The frequent shifting of the channels of the Brahmaputra causes sand accumulation in certain areas rendering the lands uncultivable at least temporarily and such lands regain fertility after a few years (5-6 years), provided sand is not deposited meanwhile. The village Kaplabori does not possess any kind of fallow land.

The cultivated area covers relatively a low proportion of area in the middle plain villages (Table 3.3). Here very high population pressure on land causes transformation of the cultivated areas into non-agricultural uses. In these villages normally land is left fallow by the rich farmers who have sufficient land and income from other sources. The tribal villages with respectively less population density have the highest proportion of land under cultivation as the tribals depend mainly on agriculture for their livelihood.

The cultivable wasteland constitutes very low proportion of area in all the villages except Kaplabori. A large proportion of cultivated land of Kaplabori was transformed into wasteland by the process of bank erosion and sand deposition of the Brahmaputra in the recent past. In the middle plain villages almost all economically viable wastelands have been brought under the plough by now. Almost so is the case in respect of the northern high plain villages. Thus, the existence of wasteland in the middle and northern high plain villages is found to be negligible.

Landholding Size and Fragmented Plots

Determined by the physical condition, history of settlement and social set-up, the size and distribution of land-holdings in the villages under this study exhibit different patterns. Table 3.4 presents the proportion of area in different land-holding size-classes.

The average size of land-holding varies between 1.22 and 2.54 ha, Kaithalkuchi and Kaplabori having the lowest and the highest respectively. Kaithalkuchi is followed by Amayapur with an average size of 1.37 ha, while for Dalbari and Kachubari the figures stand at 1.78 and 1.63 ha respectively. It is thus evident that the average land-holding size in the middle plain villages is respectively small. The size is found to be bigger in the southern floodplains as well as northern high plain villages.

The proportion of area in different size classes also varies greatly from village to village (Table 3.4). In Kaplabori 78.33 per cent of the area is in the category below 4 ha size, the minimum land required for subsistence of a family at the existing level of technology. The area in the category of less than 2.0 ha constitutes 44.23 per cent of the total area operated. These figures for Amayapur stand at 81.82 and 34.10 per cent respectively. In Dalbari all the land-holdings are of below 4 ha size and the percentage of area below 2 ha size is 31.27. The highest proportion of area under big land-holding (> 4 ha) occurs in Borajol (30.33%) followed by Kaplabori (21.67%).

TABLE 3.4: Percentage of area in different landholding size-classes

Village	Average size of landhol-ding (in ha)	% of area in landholding size-classes			
		< 1.0 ha	1.0-2.0 ha	2.0-4.0 ha	> 4.0 ha
Kaplabori	2.54	12.87	31.36	34.10	21.67
Amayapur	1.37	13.08	21.02	47.72	18.18
Kaithalkuchi	1.22	10.01	36.72	58.87	12.42
Borajol	1.41	14.49	16.83	38.40	30.33
Dalbari	1.78	8.98	22.29	68.73	0
Kachubari	1.63	11.02	27.45	56.08	5.45

Source: Field work, 1985-86.

It is to be noted that Dalbari presents somewhat a better picture in respect of distribution of area under different size-classes. As high as 68.73 per cent of the village falls in the 2-4 ha size class. On the other hand, in Borajol the percentage of area below 2 ha size (31.27%) is almost equal to that of the area above 4 ha size (30.33%). The area in 2-4 ha size accounts for 38.40 per cent in the village.

Size of Agriculture Plots and their Scattering

The size and number of agricultural plots are determined by the degree of fragmentation of holdings, resulting from the law of inheritance and the frequent transfer of land. Generally the fragments of different sizes are scattered with varying distances from the peasants' homes (Table 3.5).

The agricultural plots are scattered and located at a longer distance from the residence of the peasants in the case of the Kaplabori village. Being affected by flood and bank erosion of the Brahmaputra, many peasants of this village migrated from the Bhelakhati area on the south at a distance of about 3 km.) from Kaplabori. So they still possess medium (0.42 ha) in this village.

In the middle plain villages, the average size of plot is very small (0.29-0.4 ha). This is due to immense population pressure on land and consequent fragmentation of holdings. The average distance from the peasants' home to the plots (0.36-0.54 km.) is relatively less in these villages. In the high plain tribal villages on the other hand, the average plot size is bigger and the distance is neither very high nor very low (0.40-0.46 km.). It is thus seen that the average plot-size in the middle plain villages is smaller than that of the villages in the active floodplain and the northern high plain zones. It is worth mentioning here that no large single-plot-holding has been observed in the villages under study. The nature of scattering of the maximum fragmented land-holdings of three different villages representing the ecological-social spaces of the district has diagramatically been represented in Fig. 3.2.

TABLE 3.5: Average size and number of plots and their distance from the peasants' homes

Village	Average plot size (in ha)	Average number of plots per holding	Average distance from home (in km)
Kaplabori	0.42	4	1.85
Amayapur	0.32	5	0.36
Kaithalkuchi	0.29	6	0.42
Borajol	0.40	4	0.54
Dalbari	0.58	3	0.40
Kachubari	0.47	3	0.46

Source: Field work, 1985-86.

The cultivation of distantly located non continuous small lands is difficult from the view points of time spent in transit and transit-related activities, protection and care of crops. It is observed that even the interested peasants also cannot cultivate vegetables and other *rabi* crops because of dearth of suitable land near their homesteads. Owing to the difficulties of crop protection from animals, the individual small plots located at distances beyond the control of the peasants remain vacant particularly in the *rabi* season. The *kharif* crops, on the other hand, do not need much protection from individuals as the entire production-territory of the villages remain continuously under crops.

Major Crops : Area and Yield

The area and yield of crops notably vary from village to village in response to the ecologico-social spaces (Table 3.6).

In all the villages, rice (winter rice — *sali* and autumn rice — *ahu*) has been the dominant crop. Its area varies between 63.22 and 93.92 per cent of the total cropping area of the villages, Kaplabori and Kachubari having the lowest and the highest respectively. In Kaplabori, *sali* is regularly affected by flood and so relatively low proportion of area is devoted to it. Here *bao* (deep water rice) is cultivated extensively as it can withstand the flood water. On the other

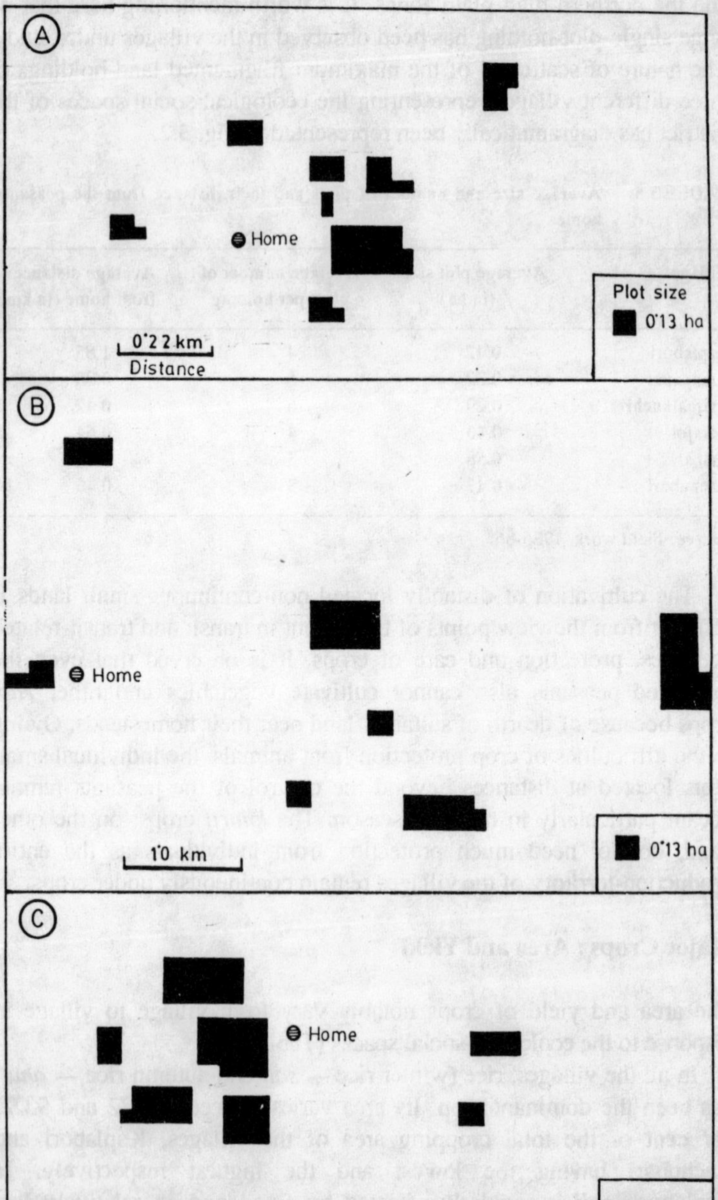

FIG. 3.2: **Size and distribution of plots of the maximum fragmented holdings in Amayapur (A), Kaplabor (B) and Dalbari (C).**

TABLE 3.6: Area and yield of major crops

Village	Area as percentage of total cropped area and yield per ha in kg.							
	Ahu	Sali	Pulses	Mustard	Jute	Potato	Wheat	Vegetables
Kaplabori	34.42 (1,287)	28.8 (1,812)	14.20 (1,387)	1.09 (617)	3.82 (2,220)	1.44 (4,162)	3.27 (1,872)	2.29 (Rs. 2,500)
Amayapur	20.05 (1,110)	65.91 (1,979)	1.54 (751)	0.30 (402)	1.65 (1,860)	1.82 (5,550)	0.13 (1,562)	4.14 (Rs. 2,100)
Kaithalkuchi	29.32 (1,093)	60.14 (1,807)	2.03 (845)	1.32 (426)	0.28 (1,829)	1.37 (3,220)	0.08 (1,478)	0.72 (Rs. 1,450)
Borajol	26.36 (1,022)	58.94 (1,842)	3.70 (1,032)	2.05 (503)	1.72 (1,931)	1.71 (5,145)	1.24 (1,482)	0.68 (Rs. 1,500)
Dalbari	23.90 (971)	58.54 (2,061)	5.19 (1,268)	4.18 (555)	0.91 (1,742)	2.84 (16,6650)	0.50 (1,779)	1.64 (Rs. 1,900)
Kachubari	31.21 (1,209)	56.71 (1,942)	2.80 (971)	0.72 (467)	0.12 (1,703)	2.14 (8,187)	0.43 (1,665)	1.06 (Rs. 1,650)

Sources: Field work, 1985-86.

Note: Figures within brackets indicate yield per hectare in kg., except for vegetables, the yield of which is given in terms of Rupees per bigha (1 bigha = 0.133 hectare).

hand, *ahu* which is harvested before the peak of summer, occupies the highest proportion of the cropped area of the village. In each of the middle plain and the northern high plain villages, *sali* dominates the cropland covering more than 58.0 per cent of the total cropped area. The per ha yield of *ahu* is found to be the highest in Kaplabori (1,287 kg), followed by Kachubari (1,209 kg). In the case of *sali*, Dalbari exhibits the highest per ha yield (2,061 kg) followed by Amayapur (1,979 kg).

Next to rice, pulses are the important crops, the area of which varies between 1.64 and 14.20 per cent. Kaplabori has the highest proportion area under pulses which include mainly *masur* (lentil), *matimah* (black gram) and *khesari*. The riverine alluvial soil of the village is very suitable for growing pulses and here their per ha yield is the highest (1,387 kg). But in the middle plain villages, pulses occupy less than 5 per cent of the total cropped area. During 1950s before the construction of embankments protecting, for instance, the village Amayapur from the flood of the Pagladiya river, the silty soil helped sufficient production of pulses without the use of fertilisers. Moreover, earlier pulses were raised by the peasants collectively (8-12 peasants together) in considerably large areas which made protection of crop much easier. But now such collective efforts are no longer there mainly for which the area under pulses has been decreasing. Almost similar is the condition in other two middle plain villages — Kaithalkuchi and Borajol. The position of Dalbari is found to be slightly better in this respect. Here the tribal peasants cultivate pulses in the Morapagladiya floodplains which being free of embankments gets fertilised naturally and become suitable for raising pulses. The tribal peasants of the village also use chemical fertiliser in their agriculture as they get it from the government at a subsidised rate. Thus the productivity of pulses is at a high level (1,268 kg) in Dalbari — compared to the middle plain villages.

Mustard, another important *rabi* crop, is cultivated in all the villages but with varying proportions of area. While Dalbari exhibits the highest proportion of area (4.58%), Kaplabori shows the highest per ha yield (617 kg) of mustard. Its proportion is relatively low in Kaplabori, because of extensive cultivation of *son* in the areas suitable for mustard. *Son* covers 13.0 per cent of the total cropped area of the village. The immigrant peasants, whatever may be their size of land-holding, generally cultivate *son* as it is easier to cultivate and it can fetch more cash per unit-area than even the rice. The main factor behind the cultivation of *son* is its special capability of fixing enough of nitrogen in the soil which enhances the yield of the subsequent *kharif* crop considerably. Except the immigrant-dominated Brahmaputra riverine tract, *son* is cultivated nowhere in the district.

Jute assumes importance only in Kaplabori. In the remaining villages, its cultivation is insignificant. The immigrant Muslim peasants cultivate it as a cash crop, while the tribals as well as the non-tribals cultivate it in minuscular plots just to meet their domestic requirements.

Potato is more or less cultivated in all the ecological zones by all the communities. But it is cultivated with full care and importance only in the tribal-dominated northern belt of the district. For instance, in Dalbari potato covers 2.84 per cent of the total cropped area and yields 16,650 kg per ha. In the villages of the middle plain and the active floodplain, the proportion of its area does not exceed 1.82 per cent. The per ha productivity of potato in these villages is less than half of that of Dalbari. The general opinion of the tribal peasants of Dalbari in connection with potato cultivation seems to be worth mentioning here. While questioned regarding the productivity of potato, Neula Narzary, a tribal peasant possessing a land-holding of medium size, but bearing a progressive outlook said,

> "We produce big potatoes,
> but get only small coins.
> We purchase seeds at Rs. 4-5 per kg,
> but have to sell the produce just after
> the harvest at almost half of a rupee
> per kg. Let us have proper storage
> facility, we shall feed the whole district."

Wheat is a late comer to the agricultural scenario of the district. Its cultivation is still very scanty and is, by and large, confined to Muslim immigrant-inhabited areas. Kaplabori has the highest proportion of area (3.27%) and highest yield per hectare (1,872 kg) of wheat. In Amayapur and Kaithalkuchi, its cultivation is almost absent. It is observed that mainly for want of irrigation on the one hand and scarcity of threshing machines on the other, the peasants are unwilling to cultivate this crop.

Vegetables which mainly include cabbage, brinjal, radish, tomato, ribbed gourd, lady's finger, cucumber, cow pea, been, sweet gourd, bottle gourd, onion and ginger, etc. although cover small proportion of area, are really of great importance in terms of money they normally fetch. It is, therefore, meaningful to assess the productivity of these crops in terms of money rather than weight. While investigated, the peasants also gave a rough estimate of their vegetable produce in money-value per unit area (Table 3.6).

The amount of money the peasants can earn from the sale of their vegetable produce per unit-area depends upon several factors such as type of crop, proximity to market, transport facility, cropping intensity, etc. It is evident from Table 3.6 that the per hectare yield of vegetables is the highest in Kaplabori followed closely by Amayapur. In Kaplabori, several types of vegetables are grown from the land through the usual practice of intercropping. Thus, although the price per unit of produce is lesser here due to mainly its remote location from the market, the total amount of money earned per unit area remains higher. On the other hand, the price per unit of produce is higher in Amayapur, because of its proximity to the urban market of the Nalbari Town. But compared to Kaplabori, the total amount of money earned per unit-area is slightly low here because of relatively low intensity of vegetable cropping. It is important to note that the peasants of Kaplabori are forced to cultivate a variety of vegetables in the *rabi* season in order to compensate the *kharif* crop damage caused by flood. In the northern high plain area, the yield of vegetables is found to be lesser due primarily to the lack of infrastructural facilities like market and transportation.

Application of Modern Inputs and Implements

The application of modern inputs such as fertiliser, irrigation, mechanical implements and adoption of HYV seeds indicate the degree of agricultural modernisation. The level of their use generally varies with the variation in the ecological condition and more significantly with that of the socio-economic set-up. However, in the family-based small farming, there is a limitation of the use of mechanical inputs due to lack of capital and smallness of land-holding size.

The representative villages having different ecological and socio-economic background exhibit notable variation in the adoption of modern technology in agriculture (Table 3.7).

TABLE 3.7: Use of modern inputs/implements and adoption of HYV rice

Village	Fertiliser input in kg/ha	% of gross irrigated area to total cropped area	% HYV rice area to total rice area	Modern implements (in number)		
				Power tiller	Sprayer	Weeder
Kaplabori	4	—	40	—	6	—
Amayapur	24	22.3	46	—	5	11
Kaithalkuchi	3	2.0	34	—	3	—
Borajol	7	4.7	30	—	—	8
Dalbari	21	—	56	1	17	39
Kachubari	12	—	49	—	9	16

Source: Field works, 1985-86.

Table 3.7 shows that the per hectare input of fertiliser (urea, super phosphate and potash) in the villages is generally low and that too varies greatly from village to village. In the tribal villages (Dalbari and Kachubari), the input of fertiliser is high. Majority of the tribal peasants apply fertiliser in their fields. It is encouraging to note that the peasants of the village Dalbari normally use fertiliser for growing vegetables and rice. The use of relatively high amount of fertiliser (2 kg per ha) by the peasants of the village is due to the government supply of it to the scheduled tribes at a subsidised rate on the one hand and the persistent effort of an efficient and devoted *gramsevak* (village level agricultural extension worker) on the other towards modernisation of agriculture in the area. In Amayapur also, the impact of fertiliser is relatively high (24 kg per ha). Here fertiliser is mainly used for growing vegetables which have a great demand in the nearby urban market. In the other two middle plain villages, i.e., Kaithalkuchi and Borajol, the application of fertiliser is at its minimum. In Kaplabori also only some amount of fertiliser (4 kg per ha) is used. It is of interest to note that the Muslim immigrants of Kaplabori generally give more stress on the application of human labour and skill than on modern inputs. For instance, uprooting of weeds particularly from the rice fields which is done with the help of a traditional implement called *khanti* requiring intensive care and strenuous labour, is undertaken nowhere in the district except the Muslim immigrants-inhabited areas. Moreover, the use of other very simple but useful traditional implements like *bidha* (a type of plough with 14-18 wooden or bamboo-made teeth at the base) by the immigrant peasants for proper spacing of the broadcast rice is unique and of great significance so far as the productivity of the broadcast *ahu* is particularly concerned.

In the district as a whole the provision of irrigation is very poor. The gross irrigated area in the district constitutes only 5 per cent of the gross cropped area. The lift irrigation projects and the deep tube wells installed by the Irrigation Department and the shallow filters installed by some of the rich peasants are the main sources of irrigation. Among all the representative villages, only in Amayapur there are two deep tube wells which irrigate 26 per cent of the village's total cropped area. Here irrigation is used only for growing rice. On the other hand, in the northern high plain villages, there is no modern irrigation facility; but the man-made *dongs* (rivulets) are traditionally used for irrigation. In Kaithalkuchi and Borajol, the naturally available water normally supports the crops. Due to the lack of irrigation, cultivation is difficult particularly in the relative higher fields, when there is no or very little rainfall.

The cultivation of HYV crops is closely associated with the use of irrigation and fertiliser. In the case of rice, Aijong, a HYV rice, has been cultivated extensively throughout the district. More than three-fourths of the HYV rice area is occupied by Aijong alone. The proportion of HYV rice area is found to be higher in the tribal-dominated high plain villages. The immigrant peasants of the floodplain on the other hand, are not much interested in cultivating HYV rice. Interestingly, some of the traditional varieties like *Hasan Gaonburha*, *Kala Manik* and *Ghuni* cultivated by the immigrants yield higher than the HYV rice. In the middle plain villages except Amayapur, the proportion of area under HYV rice is relatively low mainly because of dearth of adequate and timely supply of water. Moreover, the peasants having very small size of land-holdings do not like to take the risk of adopting a new variety immediately.

So far as the use of mechanical implements is concerned, the high plain villages are at a better position (Table 3.7). In Dalbari almost every family has at least one weeder and for every 4-5 families there is a sprayer. In the middle plain and active floodplain villages also some of the rich peasants have their own sprayers. Presently, the sprayers have become highly essential for applying insecticides especially to fight with the rice hispas *(Dicladispa armigera)* which have been causing considerable damage to rice for the last few years. It is very important to note that only one family has a power tiller among all the households of the villages under this study. Thus, it can be generalised that the wooden plough with an iron mould-board drawn by a pair of bullocks (cows are also used sometimes by the immigrant Muslim peasants) continues to be the leading implement used in agricultural operation in all the areas of the district irrespective of ecological and social conditions.

Conclusion

This study based on personal field work done in six villages representing the broad ecological settings and the major human communities of the Nalbari District points to the fact that the pattern of agricultural performance varies significantly with the variations in the ecological factors and human characteristics. The three major ecologico-social spaces, viz., immigrant-dominated floodplain, indigenous non-tribal-dominated middle plain and indigenous tribal-dominated high plain, differentiate the performance of agriculture in terms of cropping patterns, crop yield and application of modern technology.

It is evident that the active floodplain primarily under the occupance of the traditionally efficient immigrant peasants is at a higher level in the total performance of agriculture. In the socio-economically backward northern high plain area, where tribal culture predominates, the agricultural performance is at the medium level. On the other hand, agriculture is, by and large, taken half-heartedly by the relatively advanced indigenous non-tribal people of the middle plain with high literacy level and greater opportunity of earning livelihood from occupations other than agriculture. Therefore, except the small irrigated patches, the middle plain in general exhibits a low level of agricultural performance. Thus the study in the context of the Nalbari District negates the common presumption that the tribal peasants are less innovative compared to their indigenous non-tribal counterpart in the field of agriculture.

4

ENVIRONMENTAL CONSTRAINTS ON AGRICULTURE IN A COLD DESERT

HARJIT SINGH

THE STUDY of man-nature interaction is the essence in geography. It is the level and the nature of this interaction that gets reflected in the economy of a region. Man, unlike animal, does not only adapt himself to nature but interacts with it, modifies and transforms it and gets transformed in the process. This relationship is of a dynamic nature. Nature is the determining force in the early stages of development where man essentially accommodates himself within the limits laid down by it. As development proceeds onward subservience to nature gets progressively reduced and the process of cooperation and of interactions begins. The response of man towards nature is generally *region specific* depending on the resource base, the level and the nature of the natural constraints and that of technology available. For example, if the region is mountainous in terrain with cold arid climate and inhabited by simple societies, human response will be different from that in a plain area with a favourable climate. Man in such mountainous regions generally adapts himself to the requirements of nature making only limited modifications in the latter. Man responds to nature generally by adopting a pastoral economy or by carrying on a low-level subsistence farming depending on the degree of environmental severity and the level of technology available. Ladakh is a typical of such areas where nature expresses its hostility through mountainous terrain and cold arid climate, and the technology available is of low order. The degree of harshness of nature is not uniform within Ladakh, the response of man, therefore, varies considerably. The region can be divided into two parts on the basis of human response to inclement environment:

(i) The pastoral areas situated in the higher parts of Ladakh especially in its eastern part called Changthang. Excessively cold and arid climate and absence of natural drainage are the main constraints in the area. Almost entire economy of Changthang depends on natural pastures which are limited and scanty; and it forces man to keep wandering along with his herd in search of fodder. Agricultural use of land is not possible at the present level of technological development.

(ii) The agricultural areas which are confined to the lower parts of Ladakh and correspond with the river valleys. Farming is largely of subsistence type practised on areas situated below 4,500 metres above the mean sea level.

This study is concerned with the second part of the region. The purpose is to understand the extent and nature of environmental constraints on the agricultural economy of Ladakh and to have an idea about the human response to these constraints therein (Fig. 4.1). Ladakh, the extreme eastern part of Jammu and Kashmir State, occupies approximately the grid from 32°15′ N to 36° N Latitude and 75°15′ E to 80° E Longitude. Covering an area of 9,5876 sq.km., it consists of two districts — Kargil and Leh — of the State. There are 235 inhabited villages and two towns in the region. Ladakh, situated as it is in the highland crystalline mass of trans-Himalayas faces problems in the way of agricultural development which are quite specific to it. The following main natural constraints affect the agricultural economy of Ladakh:

(i) Mountainous terrain that restricts the land available for cultivation. Absence of soil cover, as large tracts of land are either devoid of soil cover or it is too thin to sustain crops, mainly due to steep slopes and other topographical factors;

(ii) Scarcity of water as precipitation is very low; and

(iii) High altitude that considerably lowers the temperature.

In general, the entire region is an elevated territory presenting a system of alternating valleys and mountain ranges except in the eastern part which is a plateau area. The mountain ranges form the most important physiographic feature of Ladakh in terms of areal coverage. Though presenting a positive physical relief, these ranges are almost totally negative from the point of view of agricultural activity. The ranges, running nearly parallel to each other, traverse Ladakh with a south-east to north-west trend. The following are the important ranges of Ladakh:

(i) The Zanskar Range,
(ii) The Ladakh Range,
(iii) Northern face of the Great Himalayan Range, and
(iv) The Soltaro Ranges and the Southern face of the Karakoram Range.

All these ranges provide barren, rocky surfaces largely devoid of soil and vegetation cover as well as of human settlements. The eastern part of Ladakh, i.e. Changterang plateau, may also be regarded largely negative from agricultural point of view. the terrain and climatic constraints do not favour settled agriculture. The economy, therefore, is largely nomadic. The nomads, called Changpas in this region, keep wandering in search of grazing lands which occur in the valleys of small rivulets or near the lakes. The plateau region is less rugged compared to the mountains and is "wonderfully even", yet it is of little use for agricultural activity mainly because of:

(i) High latitude (generally above 4,500 metres) which makes climate inhospitable. The low temperature conditions do not permit the crops to grow to maturity;
(ii) Aridity of land with an inland drainage provides a limited and highly uncertain water supply for irrigation;
(iii) The extensive rocky surfaces either with very thin soil cover or none at all.

Because of the factors discussed above, nearly 68 per cent of the total land (Fig. 4.1) lies above 5,000 metres above the sea level and is virtually unfit for vegetative and human life. The land lying between 4,500 metres and 5,000 metres, constituting approximately 5.58 per cent of the total area, permits pastoral activity of a rudimentary kind. Settled agriculture is confined to areas up to 4,500 metres in height. As a consequence of an extensive area being covered by uninhabitable mountains and barren plateaus, only 577.16 sq.km. or 0.6 per cent of the total geographical area is inhabited. Even within the assessed area*

* Assessed area is the area covered by villages as per village records.

of 577.16 sq.km., only 28.23 per cent of the land is under cultivation. This is mainly due to thinness of the soil cover and acute scarcity of water. It is only on the valley floors that one finds a relatively thicker soil cover which can support the plant life. In the light of the above, most of the level land required for agricultural activity is found associated with the following geomorphic features:

(i) The velley floors mainly on the river terraces where land is level with a reasonably thick soil cover, and
(ii) Alluvial fans and talus-cones where slopes are gentler and land can be cut into terraces.

Since rainfall is very low and farming without irrigation is not feasible, availability of water is another prerequisite for the sustenance of agricultural activity. Consequently, all the agricultural land is situated in close proximity to perennial streams or springs. Though the river valleys occupy a very small proportion of the region in terms of areal extent they have a great significance in the agricultural geography of Ladakh. It becomes clear from the above discussion that agricultural geography of Ladakh is mainly concerned with the interaction of man with environment within the river valleys of the region which provide the bulk of the cultivable and habitable land in an otherwise negative area.

The second major constraint on agriculture in Ladakh is imposed by cold and arid high-altitude climate. In such areas of hostile climate inhabited by simple societies, man is largely at the receiving end and has barely emerged from the realm of necessity to that of freedom. Climate exerts a significant influence on the economy especially on its agricultural sector. Agricultural activity is carried on within the limits imposed by climatic constraints. Man endeavours to weaken the grip of these constraints with the help of primitive technology and has been successful in slightly widening the areas of economic operations under harsh climates. In the field of agriculture he generally moves in two directions to circumvent the climatic constraints.

First, by overcoming the deficiencies created by climate in a particular area with the help of artificial means, e.g. natural scarcity of moisture required for crops is counteracted with the help of irrigation.

Secondly, by modifying agricultural practices so as to correspond with the environmental limits, e.g. by introducing early maturing varieties in areas where temperature is very low, over a longer period of year or by cultivating crops requiring less water in the arid zones.

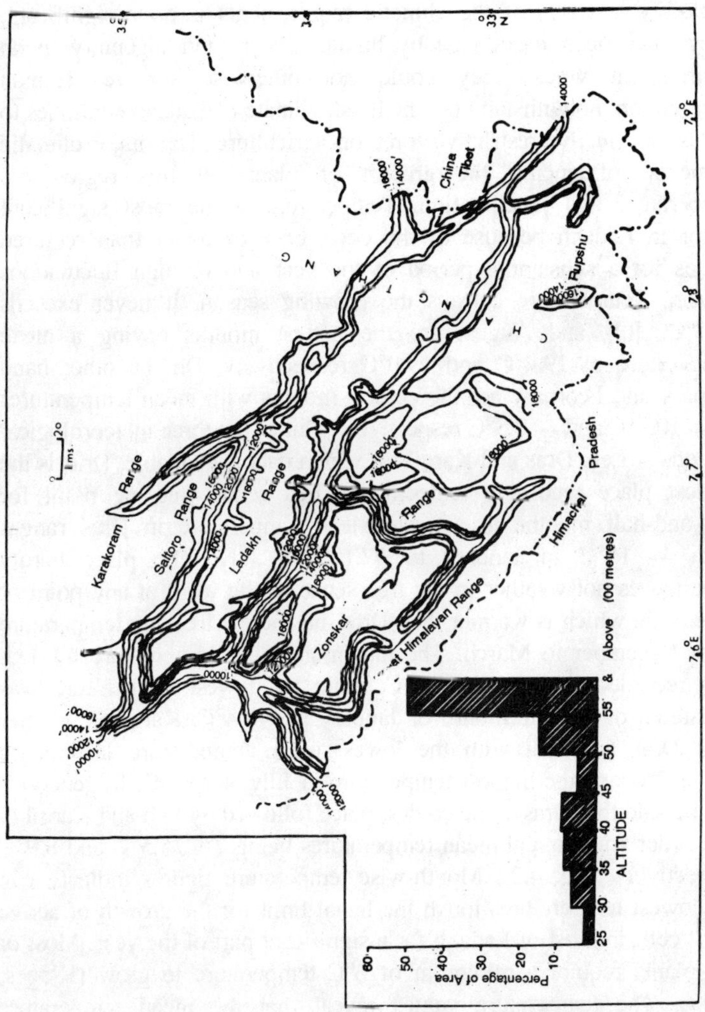

FIG. 4.1: Ladakh: area in different height groups.

Most of the plants have their optimum climatic requirements. Unlike natural vegetation, agricultural crops have been domesticated and modified from their original state; and, consequently, their flexibility in terms of the climatic requirements is quite significant. They .have been introduced by human efforts and ingenuity in an environment where they could not otherwise survive. Human intervention, notwithstanding, the harsh climate of Ladakh continues to act as an effective restrictive force on agriculture. The major climatic elements influencing the growth of plants in this region are temperature and precipitation. Temperature is the most significant factor in Ladakh because of the occurrence of lower than required values for a substantial period of the year and of high fluctuations therein. Temperature curtails the growing season. It never exceeds 29.7°C. July and August are the hottest months having a mean temperature of 19.4°C and 19.0°C respectively. On the other hand January and February are the coldest months with mean temperatures of — 10.9°C and — 8.5°C respectively. Out of the three meteorological stations at Leh, Dras and Kargil for which data is available, Dras is the coldest place recording temperatures below the freezing point for five-and-half months in a year. Mean temperature in Dras ranges from — 15°C in January to 17.1°C in July. The place hardly experiences hot weather in the true sense of the word at any point of time. Leh, which is warmer than Dras, has below freezing temperature from December to March. The maximum mean temperature for Leh was recorded in July at 17.4°C while the lowest temperature was registered during the month of January at —8.4°C. Kargil is warmer than Dras and Leh with the lowest mean temperature in January at —8.7°C and the highest temperature in July at 23.7°C. In general it can be said that Dras is the coldest place followed by Leh and Kargil in that order with annual mean temperatures being 2°C, 5.5°C and 8.9°C respectively (Fig. 4.2). Month-wise temperature figures indicate that the lowest temperatures touch the lethal limit for the growth of active plant cells in most of Ladakh for a significant part of the year. Most of the plants require a minimum of 6°C temperature to grow (Klages, 1948). The temperature values reveal that the mean temperature remains below this critical value for nearly half of the year in Leh and Dras and for one-third of the year in the case of Kargil. The duration of the growing season for each of the three stations was found to be as given in Table 4.1.

TABLE 4.1: Growing season (based on mean temperature)

Station	Height (metres)	Growing Season	
Dras	3,066	May 16 to	October 14
Leh	3,514	April 18 to	October 18
Kargil	2,682	April 8 to	November 22

Though temperature is directly linked with altitude, Dras is an exception due to its location in a funnel-shaped valley facing Zojila in the shadow of the Great Himalayan range. The place, therefore, receives a higher amount of snowfall and rarely faces the sun, thus, temperature remains below 6°C for a longer time entirely due to local relief factors.

The growing season is even further reduced due to the high diurnal range of temperature. Though the mean temperature may be higher than 6°C but, quite often, the minimum temperature stays below this value. For example, although the mean monthly temperature for Leh in the sowing month, i.e. in May, is 9.9°C but because of a high diurnal range it goes below the critical value quite a number of times during the months (Fig. 4.3). In May of 1965, for example, the temperature crossed the critical value on 28 days and even went below freezing point on 10 days. Such a high diurnal range, therefore, damages the crops. In response to these daily fluctuations in temperature, the sowing seasons get delayed and generally begin in the beginning of May in Kargil, in the second half of May in Leh and in the beginning of June in Dras. The harvesting season also gets staggered from the beginning of September to the end of October depending on local climate conditions (Fig. 4.2).

Further, it is not only the number of calendar days of the growing season which is important, but also the total energy units available to the plants during this period. In order to assess the available energy units, "degree days" have been calculated. A degree day for a given crop is defined as a day on which the mean temperature is one degree above the "zero temperature" (Critchfield, 1968) which has been taken as 6°C in this case.

Table 4.2 reveals that maximum degree days are available at Kargil followed by Leh and Dras. It is clear that not only is the growing season longer at Kargil but it is at the same time more effective.

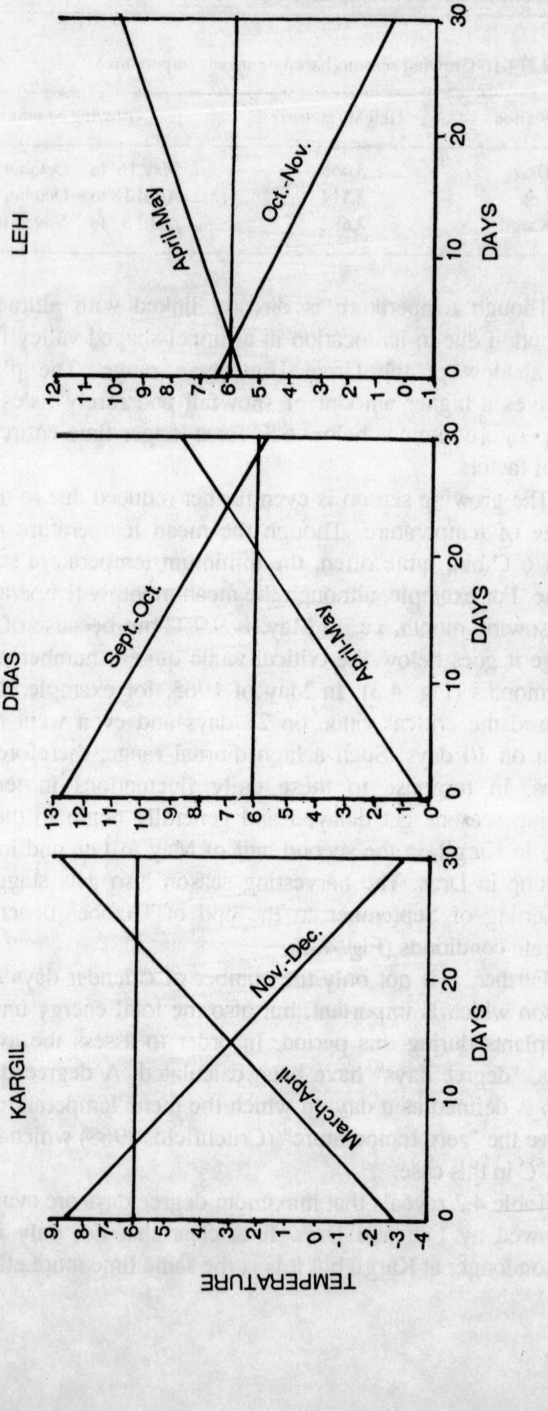

FIG. 4.2: Growing season.

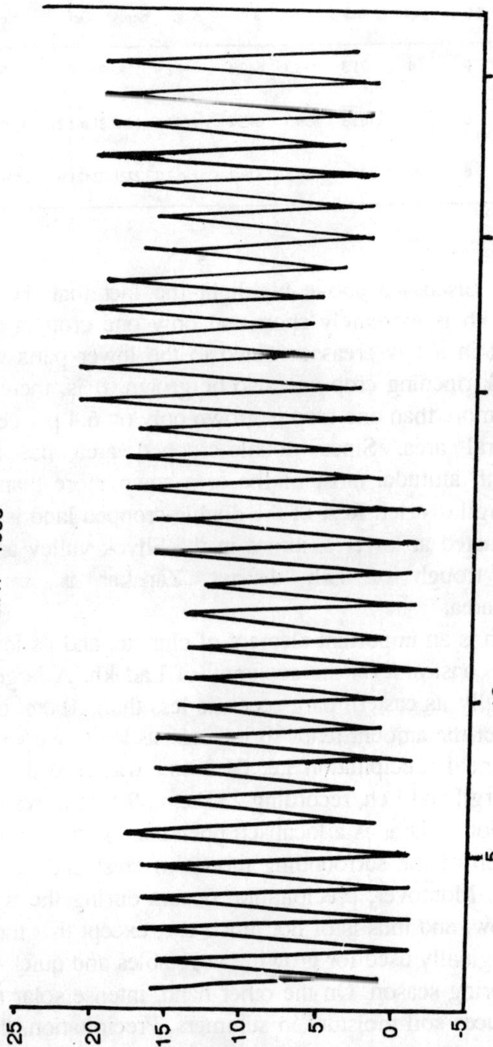

FIG. 4.3: Leh — diurnal range of temperature.

TABLE 4.2: **Accumulated degree days**

	Jan.	Feb.	March	April	May	June	July	Aug.	Sep.	Oct.	Nov.	Dec.
Dras	0	0	0	0	71.3	296.3	640.4	982.9	1183.9	0	0	0
Leh	0	0	0	0	102.3	339.3	692.7	1030.6	1243.2	1262.2	0	0
Kargil	0	0	0	81.0	375.5	780.5	1329.2	1857.7	2238.7	2421.6	2496.6	0

The factors discussed above highlight the fact that the growing season in Ladakh is extremely short and only one crop in a year is possible except in a few areas confined to the lower parts where an additional quick-ripening crop can also be grown. It is, therefore, not surprising that more than one crop is grown only on 6.4 per cent of the total agricultural area. Since double-cropped area has a direct relationship with altitude, most of the area sown more than once is confined to Kargil district. Rest of the double-cropped land is found in the villages situated at lower altitudes in the Shyok valley and in the Lower Indus trough of Leh district. Zanskar is an entirely single-cropped area.

Precipitation is an important element of climate, and its low values act as a strong constraint on the economy of Ladakh. A large area of Ladakh, especially its eastern part, receives less than 10 cm. of annual precipitation, but the amount tends to increase as we move westwards. The highest annual precipitation i.e. 64.8 cm. was recorded at Dras followed by Kargil and Leh, recording 23.9 cm., 9.14 cm. respectively. High precipitation at Dras is a localised phenomenon occurring due to the configuration of the surrounding mountain crest and its situation opposite Zojila. Moreover, precipitation occurs during the winters in the form of snow, and thus is of not much use, except that the ground moisture is marginally used for growing vegetables and quick-ripening crops during spring season. On the other hand, intense solar radiation immensely reduces soil moisture in summers. Precipitation, therefore, has a negligible direct role to play in the agricultural economy of Ladakh. The water deficiency is largely met with the help of irrigation. Stream and spring water is used for irrigation through the indigenous system of *Kuls*.

In order to measure the extent of water deficiency the Thornthwaite method has been used and potential evapotranspiration (P/E) is calculated.* Though the method has rather limited validity for a region like Ladakh because of inherent inadequacies, it does provide a rough idea about the gap that exists in the demand and supply of water. The major limitations of the method in terms of its applicability in Ladakh are as follows:

First, it consider evapotranspiration to be zero when temperature is below freezing point. It neither takes into consideration the evaporation from the snow surface nor of that caused by wind.

Second, it assumes vegetation cover on the surface which is largely non-existent in Ladakh.

Finally, the method is based on mean temperature and precipitation, and does not take the effect of diurnal range into account. Table 4.3 gives the precipitation and P/E (in inches) for three stations.

Table 4.3 and Fig. 4.4 reveal that the moisture requirements are much greater than the availability during the growing season. Since the potential evapotranspiration is primarily a function of heat; it can be seen that it is greater at Kargil compared to Leh and Dras. Moisture is in surplus during winter months when temperature is below freezing point; as a consequence it has no direct utility for agriculture. Since the scarcity of moisture is very acute from April to October, nearly 100 per cent of the net sown area is under irrigation. It is only double-cropped area where mainly vegetables and *Trumba* (Buck wheat) are grown without irrigation. These crops utilise moisture retained by soils from the winter snow. Besides, a few crops are grown without irrigation in the wetter valleys of western Ladakh, the area occupied by them is negligible. Scarcity of water effectively restricts land under cultivation. It was observed that nearly 15.25 per cent of the assessed land designated as "culturable waste" can be brought under plough if irrigation facility can be extended to this land.

---

*

$$P/E = 11.3\left(\frac{P}{t-10}\right)\frac{10}{9}.$$

P/E = Potential Evapotranspiration.

P = Precipitation.

t = mean monthly temperature.

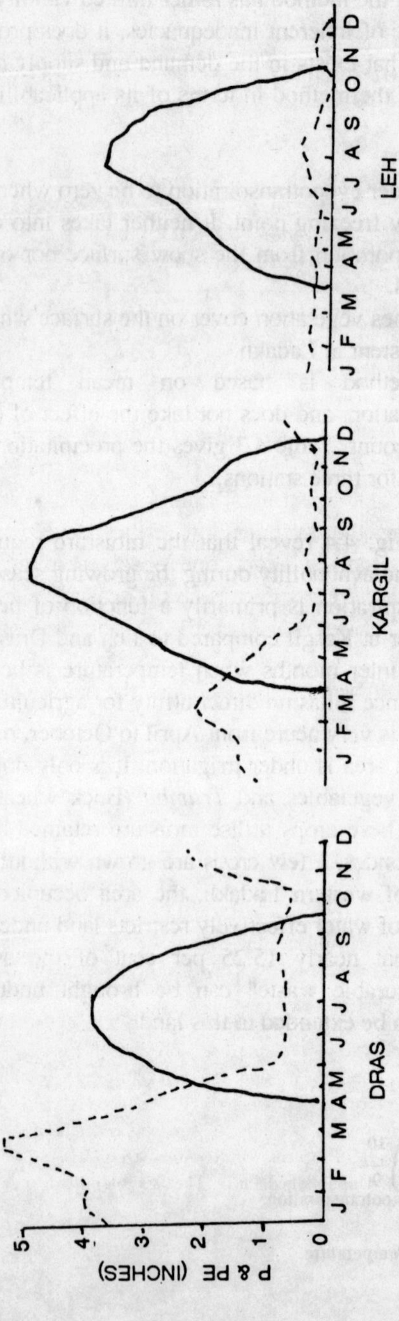

P = Precipitation

PE = Potential Evapotranspiration

FIG. 4.4: Moisture deficiency.

TABLE 4.3: Precipitation and potential evapotranspiration

(*in inches*)

		Jan.	Feb.	Mar.	April	May	June	July	Aug.	Sep.	Oct.	Nov.	Dec.
Dras	Pre	3.8	3.8	5.4	4.09	2.42	0.7	0.6	0.5	0.7	0.8	0.5	2.1
	PE	0	0	0	0	2.1	3.1	3.8	3.8	2.8	1.4	0	0
Leh	Pre	0.4	0.3	0.3	0.2	0.2	0.2	0.47	0.6	0.3	0.1	0.4	0.2
	PE	0	0	0	1.3	2.0	2.7	3.8	3.6	2.8	1.4	0	0
Kargil	Pre	1.3	1.3	2.1	1.7	1.0	0.2	0.3	0.4	0.1	0.3	0.1	0.6
	PE	0	0	0	1.5	2.8	3.8	4.9	4.7	3.3	2.0	1.3	0

Pre = Precipitation. PE = Potential Evapotranspiration.

To summarise, inclement climate and mountainous terrain are the major constraints on the agriculture of Ladakh, climate becomes a further deterrent during the winter season. Though agriculture absorbs 80.7 per cent of the workforce, it barely meets the food requirements due to the environmental constraints. Entire agricultural economy is of subsistence type, and the avenues for further development of this activity are also limited. In short, one observes a highly seasonal economy in Ladakh in which winter is an off-season for farming. Only indoor economic activities are possible during winters which, at present, are spinning and weaving. Consequently, any development strategy for Ladakh must take into account the seasonal character of the economy.

REFERENCES

Biscoe, C. E. T., 1972 : *Kashmir in Sun Light and Shade*, Sagar Publications, New Delhi, reprint.

Boughey, Arthur S., 1971 : *Man and Environment — An Introduction to Human Ecology and Evolution*, Macmillan, New York.

Cunningham, Alexander, 1970 : *Ladak*, Sagar Publications, New Delhi, Originally Published, 1853.

Drew, F., 1972 : *Territories of Jammoo and Kashmir*, 1875, Orient Publishers, Delhi.

Francke, A. H., 1925 : *Antiquities of Indian Tibet*, Vol. I and II.

Hugel, B. C., 1972 : *Kashmir and the Punjab*, Life and Life, Delhi Reprint.

Critchfield, J., Howard, 1968 : *General Climatology*,, Prentice Hall, New Delhi.

Klages, K. H. W., 1948 : *Ecological and Crop Geography*, Macmillan, New York.

Singh, Harjit, 1978 : Ladakh: *Problems of Regional Development in the Context of Growth Point Strategy*, Unpublished Ph.D. Thesis submmitted to Jawaharlal Nahru University.

5

PHYSICAL FACTORS AFFECTING AGRICULTURE

SUBHASHRANJAN BASU

BECAUSE MAN lives on Earth he depends on the Earth. Although undoubtedly natural facts are far from explaining everything, yet slope, soil, climate, hydrographic conditions, and so forth, have very important effects and are responsible in general for the order — often much too disorderly, of human phenomena. It is essential, too, for men to recognise precisely the true reality of the natural conditions that form the setting of their lives, and to know always what precise geographical facts they will have to face. With rare and skilful flexibility man's genius adapts itself to the most varied conditions: the events that do overwhelm and paralyse him are those that are abnormal or at least unexpected.

The article as such is a tentative and speculative inquiry into cultural causation, centring its attention on the issue of physical processes in origination and in change. No attempt is made to crystallise theory, but the issue is raised in the hope of focusing attention on the problems involved.

Area of Study

Selection of the lower Bhagirathi Valley for this purpose, in between the Latitudes 23°24′ N. and 23° 40′ N. and Longitudes 88°6$^{1}/_{2}$′ E. and 88°27′ E, approximately 80 km. north of Calcutta (Fig. 5.1) was in a large measure a matter of personal familiarity with the area. However, because this particular stretch of valley is still without major artificial controls against flooding, meandering and bank erosion, it affords an ideal situation for making direct observations on the effects of a river on its flood-plain configuration and the adaptations of the human beings to this changing natural landscape.

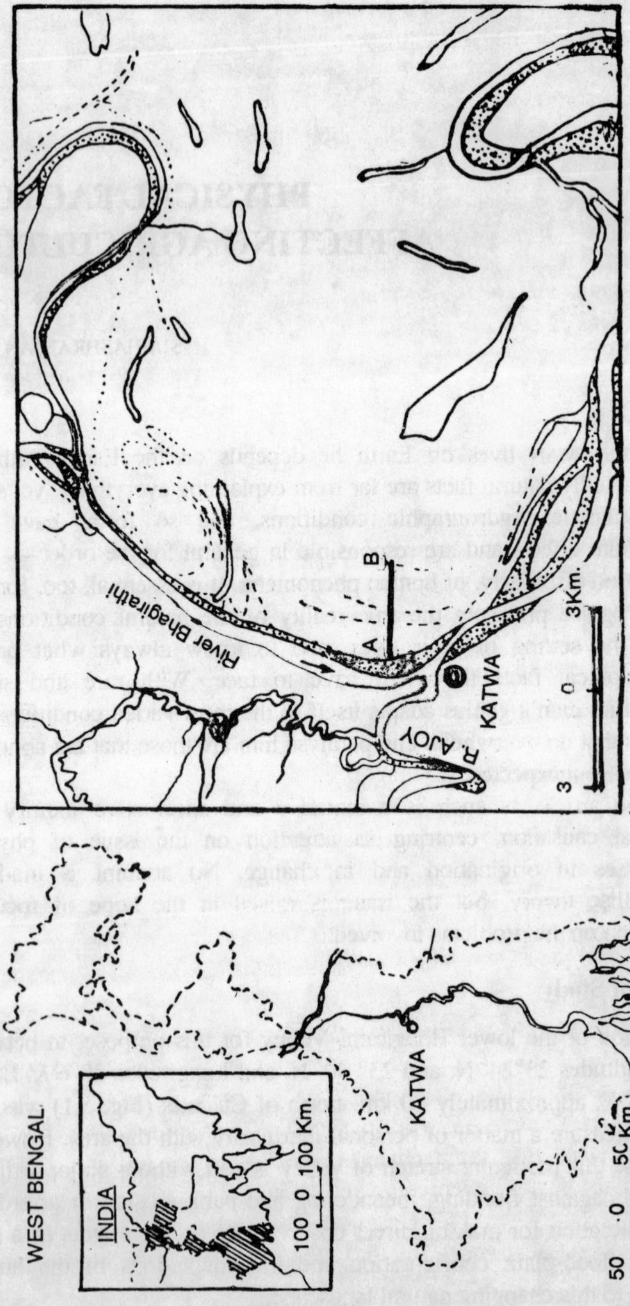

FIG. 5.1: Location of the area of study.

The description of surface configuration emphasises those characteristics that seem to have repetitive occurrences and pattern in different parts of the valley. For the most part, the pattern of configuration on individual parts of the flood-plain and the arrangements of these parts within the lower Bhagirathi Valley are provided for in the maps and illustrations.

Quantitative information on flood-plain topography and dimensional characteristics of specific portions of the flood-plain together with the landuse was mostly obtained through detailed field investigations involving several separate excursions. The other types of data required were obtained from published data and maps of The River Research Institute, West Bengal, and The Survey of India, Calcutta.

Fluvial Characteristics of the Bhagirathi Valley

The Bhagirathi river valley has been carved in a floor of alluvia having an immense depth of over 457 m. lying on a basement of basalt lava flows, presumably of the same age as the Rajmahal group volcanoes.

The general pattern of the river reflects its natural tendency to wind back and forth within the restricted confines of its flood-plain, composed of soft alluvia with an almost homogeneous texture.

As the bed-loads of the river Bhagirathi are less susceptible to transportation and more liable to deposition, unless and otherwise there is a heavy flood, the silt-burden is dropped by the river in suitable places to form numerous shoals, which are of common occurrences within the channel of the river Bhagirathi.

The formation of shoals restricts the large volume of water to move through a reduced channel, causing an asymmetry of erosion on the opposite side of the shoals due to the favourable gradient of the channel and this invariably pronounces the formation of a meander in the same direction.

In response to its meandering habit the river Bhagirathi like a pendulum oscillates from side to side and thereby changes its course from time to time.

Maps of Rennell (1765), Lloyd (1840), Survey of India (1916-17 and 1957) and the recent changes (1965-1975) as evidenced by the author himself, have been compared to find out the exact sequence of changes made by the river Bhagirathi during the past 200 years. It is quite evident that due to the friable nature of the soil, the meander-bends of the river have been continually enlarged and pushed ahead, thereby lengthening the river course. Consequently, the general slope of the river is reduced as is also the velocity of the flowing water, until the erosive power of the current is brought into equilibrium with the resistance of the soil. Formation of cut-offs in suitable circumstances and the simultaneous migration of river-bends have induced the river to swing over its entire flood-plain which, on the contrary, has thus been largely sculptured by the river itself.

The Changing Natural Landscape

In order to investigate, in detail, the evolution of micro-relief features found on the flood-plain of the Bhagirathi, the region east of Katwa town, on the left bank of the river, has been selected. Levelling was carried on with the help of a "Dumpy Level" and section was drawn on scale 1 cm-1 m (vertical) and 1 cm = 10 m (horizontal). In this particular reach, the bend migration of the river is most pronounced in the sense that here the great "Katwa-bend" has been pushed towards the west in about five successive stages and the river is on the verge of leaving its sixth one. All these five stages of bend-migration have been carefully mapped and the associated textural variations in soils have been analysed (Table 5.1). What is most interesting is that the gentle swells are clearly marked by a flight of steps (at best three) of unequal tread and rise having a distinct terrace-like formation (Fig. 5.2). This remnant of flight of terraces has its ready likeness, with the present sequence of terraces in the process of formation along the left bank of the river.

The Process of Terrace Formation

The middle reach of the present "Katwa-bend" is conspicuously divided into two separate streams by the development of a large shoal having its concave side towards the left bank of the river. In between the shoal and the left bank there lies an almost choked and decaying stream. With the passage of time, the gradually enlarged shoal and the gently sloping point-bar side of the left bank of the river will coalesce and the river will migrate towards its right-hand side to complete its sixth stage. Since 1965, the author has kept a close vigil on the spot and in the early part of 1970, it was found that during the dry season, and even in the first phase of the monsoon, the above-mentioned choked-up portion of the river does not receive any water from the adjoining channel. This, on the contrary, widens itself towards the west to compensate for the loss incurred. The landscape thus formed on the left bank will have atleast two flights of terraces, the upper one, the levee; the lower one, the former left bank (Fig. 5.2). All these processes of terrace formation with the progressive migration of river-bends have been diagrammatically sketched in the adjoining map (Fig. 5.2). Similar type of alluvial-terraces (so named as these are composed of alluvia), which are in no way related to earth movement but are solely due to river migration, are found further inland. In the oldest or Stage-1 of bend migration, in this part of the country, two distinct terraces are found. From a height of 18m, the upper terrace slopes gently to meet the lower one at an elevation of 17m. Next comes the lowest approach at an elevation of 16m and this is nothing but the former river bed. All these are readily recognisable at a distance of 1,360 m immediately away from the left bank of the river Bhagirathi. Similar terraces are found in the case of the Stage-2 and the Stage-4 having in between the Stage-3, where three distinct terraces have been located. The uppermost terrace is found at an elevation of 19 metres, the second at 18 metres and the third at 17 metres. Coalescence of the shoal with the then left bank may be responsible for the formation of the lower terrace, when the middle one is the original left bank and the uppermost one, the levee.

Textural Variations of Soils

Minute topographic variations as reflected in the forms of the terraces have characteristically different textural variations in soils as is evidenced from the mechanical analysis of soil samples collected from different heights. The highest terrace is invariably associated with loamy sand (70.3 — 76.3 per cent of sand and 16.8 — 22.8 per cent of silt), as it is the zone on the river bank where the river regularly deposits its silt burden, mainly composed of sands. The next terrace is formed of silty loam, whereas the lowest approach comprises silty clay as the clay particles have been washed down during the rains from the adjoining higher slopes.

The Adapting Cultural Landscape

The already described varied topographic forms of ups and downs in causation to their textural variations in soils have undergone three essentially different cultural uses. Water available both from atmospheric precipitation and the adjoining river Bhagirathi is the dominant factor for their specifications.

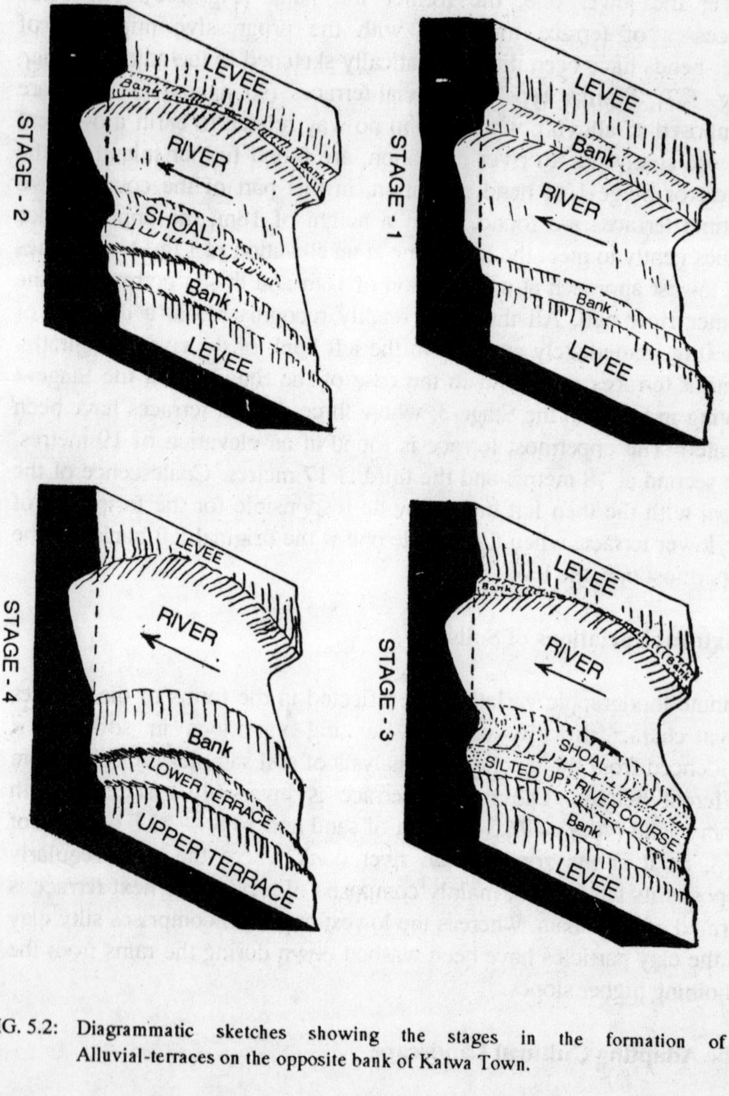

FIG. 5.2: Diagrammatic sketches showing the stages in the formation of Alluvial-terraces on the opposite bank of Katwa Town.

TABLE 5.1: Soil analysis

Sample No.	% of sand	% of silt	% of clay	Texture (According to International system)
1	72.3	18.8	8.9	Loamy Sand
2	68.9	28.8	2.3	Silty Loam
3	72.9	22.8	4.3	Loamy Sand
4	32.9	40.8	26.3	Silty Clay Loam
5	40.9	36.8	22.3	Silty Loam
6	22.9	36.8	40.3	Silty Clay
7	50.9	28.8	20.3	Silty Loam
8	72.3	20.8	6.9	Loamy Sand
9	14.3	34.8	50.9	Silty Clay
10	54.3	32.8	12.9	Silty Loam
11	76.3	16.8	6.9	Loamy Sand
12	15.6	36.8	47.6	Silty Clay
13	50.2	28.2	21.6	Silty Loam
14	72.9	20.8	6.3	Loamy Sand

N.B.: Analysed in the Soil Laboratory of the Deptt. Of Geography, Calcutta University, Calcutta.

The upper terraces of loamy sand being farthest from the occasional floods of devastating nature, both horizontally and vertically, have been well utilised for homesteads surrounded by orchards. Pulses requiring a few showers of rain and suitable to higher grounds of inferior soils are partially cultured where there is space. Cart-track for inter-village communication is also built here.

The lower-terraces of silty loam are best suited for market gardening. Inter-culture is always favoured for varied products of different values with an overall maximum per-capita income. Fertility of land is, somehow, maintained by the additions of silts during seasonal floods.

The lowest approach of silty clay is preferred for paddy culture. The high clay content coupled with the rise of water-table almost to the surface during the rains makes the area suitable for "Aman" paddy, especially its deep water variety. Effective capillary action resulting in high salt concentration in the soil does not allow a "rabi crop" to grow.

Thus, during the rains, except for the spaces required for homesteads and orchards, the whole expanse is under cultivation. But with the progress of the drought the picture differs. Instead of a green canopy there appears a desert-like greyness. As the ponds get dried up and the water in the wells reaches its lowest level, the cultivation of the terraces, further inland, is a remote possibility. The lands immediately on the river bank are cultivated only with water-melon, gourd, cummin-seed, etc., as the soils have a fairly good moisture status being nearer to the river water.

Conclusion

Thus it becomes obvious that in response to the meandering habit, the river Bhagirathi shapes its floodplain at micro-level with a simultaneous textural variation in soils which, varying in moisture status and fertility, have undergone essentially different cultural uses. Such a restricted cultural use may be improved and extended throughout the year by introducing an effective irrigation system through shallow tube-wells, but that would never alter the physical processes in origination. On the contrary, this alone would probably compel the cultivators to stick to their present agricultural practices and give them minimum option for a change in cropping pattern. Moreover, the opening of the Farraka Barrage, without fulfilling the minimum need of 40,000 cusecs of water, is yet to nullify the process of shoaling in the Bhagirathi channel which has thus been migrating freely over its vast flood-plain to characterise it with minor swells and depressions. And these by their very nature dictate terms to their agricultural users.

REFERENCES

Basu, Subhashranjan, 1968 : "Landscape Evaluation — A Methodology", *Geographical Review of India*, Calcutta, Vol. XXX, No. 4, pp. 9-13.

———, 1973 : *The Decay and the Associated Problems of River Dynamics of the Bhagirathi Valley*, unpublished Ph.D. Thesis, Calcutta University, Calcutta, Ch. 6, pp. 149-65.

———, 1976 : Some Fundamental Problems of Meander Formation with Special Reference to the Bhagirathi River, *Geographical Review of India*, Calcutta, Vol. 38, No. 1, pp. 54-67.

Basu, Subhashranjan and Kar, Nishith Ranjan, 1970 : "Meandering, Migration and Alluvial Terraces in Lower Bhagirathi Valley, West Bengal, India", *Indian Journal of Power and River Valley Development*, Calcutta, Vol. XX, No. 1, pp. 29-39.

Brunches, M. J., 1952 : *Human Geography* (abridged edition), George G. Harrap & Co. Ltd., London, pp. 212-232.

Davis, W. M., 1902 : The Terraces of the West field River, *The American Journal of Science*, Fourth Series, Vol. XIV, pp. 77-94.

Schmudd, T. H., 1963 : "Some Aspects of Landforms of the Lower Missouri River Flood Plain", *Annals of the Association of American Geographers,* Washington. Vol. 53, No. 1, pp. 60-73.

Spencer, J. E., 1963 : "How Does An Agricultural Region Originate?", *Annals of the Association of American Geographers,* Vol. 53, No. 1, pp. 74-92.

Sen, Satyakam, 1966 : "River Curves and Their Forms", Ch. IV : A Flood-plain Section, *Indian Journal of Power & River Valley Development,* Calcutta, Vol. XVI, No. 12, pp. 13-17.

Wanmali, S. H. 1983. Some Aspects of [illegible] [illegible] [illegible] [illegible] the Association of American Geographers. Washington Vol. [illegible] pp. [illegible]

Spencer, J. E. 1963. [illegible] Does an [illegible] [illegible] [illegible] of American Geographers, Vol. [illegible] pp. [illegible]

Sen, [illegible] [illegible] [illegible] [illegible] [illegible] [illegible] [illegible] Section, Indian Journal of [illegible] [illegible] [illegible] Vol. [illegible] No. [illegible] pp. [illegible]

6

GEOMORPHIC PROBLEMS OF AGRICULTURE IN ARID LANDS

AMAL KAR

AGRICULTURE IS principally based on the morphological characteristics of land on which it is practised. In fact, the land for tillage or the land for pastoral purpose has certain physical and chemical characteristics and certain limitations based on those characteristics. Man cultivates his land, sows the seeds, uses water, applies manures, fertilisers, etc., and reaps a harvest of crops, or he simply takes his domestic animals to the land for grazing the naturally regenerated vegetation and subsequently exploits the animals according to his requirements. As the pressure on land increases either through cultivation or through pastoralism, its limited hidden resources are used up at an accelerated rate and are exhausted before they could be renewed. In some cases, the use of a land in a wrong way does also accelerate the pace of deterioration. This problem of decreasing capability of land to sustain crop has been dealt with by scientists of several disciplines. Since man's use of land for agriculture does not simply mean a piece of land, but the total relief, drainage, physico-chemical constitution of the material with which the land is built, etc., in short, the geomorphic setting of the land, a geomorphologist can play, perhaps, a better role in understanding the problem and even suggest some solution to it. Because "it is with the form, material and processes of the earth's surface that geomorphology is concerned" (Cooke and Doornkamp, 1974). Despite this assertion, however, not much work on the geomorphology applied to agriculture has been done in our country.

We shall describe here briefly how the geomorphic parameters of land control the agriculture in the arid and semi-arid regions of western Rajasthan and how man's tampering with the system results in complete failure of agriculture.

Identification of the Problems

From the applied geomorphology viewpoint the three major problems of the desert and its fringe in western Rajasthan are the dearth of water, salinity of land and movement of sand (Ghose, 1977). All these three problems can be generated by nature alone, but are often accentuated by man. In other words, man has now become an important geomorphic agent. According to Ryabchikov (1975, 118) "changes caused by man's activity and transformations of the natural environment are now equivalent to natural planetary processes".

The adverse effects of man's intervention are most glaring in the delicately balanced ecosystem of the desert, because here the land cannot stand the excessive biotic stress or exploitative enterprises. Hence, the twin aspect of inherently vulnerable nature of the desert land and exploitative nature of man have to be viewed together in any rational planning of agriculture for sustained productivity.

The Problem of Water

Water is a scarce commodity in the western part of Rajasthan. Agriculture is principally based on the meagre monsoonal rainfall of less than 150 mm to about 350 mm, which is both uncertain and erratic in distribution. A large tract with sandy plains, sand dunes and other sandy undulating terrain and more problems for run-off, water conservation and even availability of adequate moisture for crops. Only a single crop of rain-fed bajra is, therefore, grown in most of these areas. To conserve the seasonal run-off, the villagers generally create embankments either at the pediment — buried pediment fringe or across the short ephemeral streams in the older alluvial plains — or they simply dig a pond where the local topography is such that water from all sides is directed towards that pond and the surface is either silty or has exposed calcified horizon. The water conserved in this process is, however, too meagre for irrigation; it is used for domestic consumption only.

In some cases the construction of such embankments across the ephemeral channels caused the channels to aggrade their beds from the tanks upstream and then shift their courses. In the process, the fields through which the new courses are cut through get coarser deposits and thereby lose productivity. Evidence of such man-induced changes in channel courses has been found around Bitan, Pundlu and Basni Sejan in the south-west of Nagaur district (Kar, 1976).

In the absence of adequate surface water resources the farmers turn to the subsurface source of water. The major source of subsurface water which has been exploited so far, lies either in the hard rocks like the Vindhyan sandstone (which is strictly geologic and will not be discussed here), or in the younger alluvial plains along the major ephemeral streams such as the Luni, the Jojri, the Jawai, and their major tributaries. As a result one finds double-cropped lands along the banks of these streams. Irrigation is done from the numerous wells.

Excepting the younger alluvial plains the other major potential source of subsurface water, which is still underexploited, is the courses of the prior drainage channels, buried under later deposits of sand and other finer particles. It has now been firmly established that this desert was not so barren all through in the past, but experienced several major cycles of dry and wet climates (Ghose et al., 1977). During the wet phases there were numerous streams flowing through the desert and there was drainage integration. Even the river Saraswati of the Vedic literature used to flow through this desert (Ghose et al., 1978). However, the drainage systems became disorganised and partially buried as and when aridity intervened the wet phases. Be that as it may, subsurface water accumulates along these buried drainage courses and can be exploited for gainful purposes. In many cases the farmers have perchance dug wells along those courses and have struck water at an average depth of 5 to 40 metres in the flat older alluvial plains and at 20 to 60 metres depth in the sandy undulating terrain. But in most other cases the water along the prior channel courses remains unutilised. Through the interpretation of aerial photographs, however, a number of prior drainage systems have now been identified and reconstructed to suggest the possible subsurface water potential zones (Ghose, 1965; *see* also the Annual Reports 1966 to 1978 of the CAZRI). For example, it has been revealed that in Merta and Degana tehsils of Nagaur district the subsurface water occurs along the prior Jojri river and its numerous tributaries, now buried under sand and alluvium but having connection mainly with the Aravalli mountains in the east. The depth of water table varies from 5 to 20 metres. Although the subsurface water along the prior Jojri is being utilised, the water along its prior tributary valleys has not yet been fully explored.

Wherever the discharge from wells is sufficient for agriculture, people are using the well water for double cropping, thus changing significantly the agricultural landscape. In the Jhanwar-Doli-Boranara sector of the Luni development block in Jodhpur district, recent exploitation of subsurface water along the prior buried channels for irrigation has already stabilised the sandy undulations of the area and has increased crop production (Singh et al., 1978). At Borunda in the south-east of Jodhpur district such recent exploitation of subsurface water has converted the previously single cropped colluvial plains into double-cropped irrigated fields. Such is the case in the alluvial plains of the Gaju-Kuchera sector of Nagaur tehsil in Nagaur district also. Still there are many more areas where the subsurface water along the prior channels remains to be exploited for gainful purposes.

The Problem of Salinity

Although the prior drainage channels of the desert can be treated as a good host of shallow subsurface water, the quality of water especially the content of salt in it, is also to be taken into account. In this regard the relation between the geomorphology of prior streams and the geochemistry has been established by Ghose (1964) in the following way: When the streams were in flow, they used to carry downstream mineral salts of various solubility along with the sediments. Subsequently when the streams' competence was lowered due to the onset of aridity, the salts began to precipitate according to their solubility. Thus calcium sulphate was deposited upstream of the regions where sodium chloride was deposited. This succession is manifest on the surface along some of the courses in the form of gypsum and commercial salt, but in others they remain underground. Therefore, when a well is dug along the course of a prior channel, caution is to be taken not to tap the underground salt sources which is generally at the confluence of the streams or in the lower reaches.

Apart from such natural cause of salinity man himself is also aggravating the salinity problem. Since the prior channels are a good host of subsurface water, stress of cultivation is more on lands wherever such water has been exploited through dug wells. But, to irrigate more and more land the farmers sometimes construct embankments across the courses of the prior channels and collect the seasonal run-off which flows along the valley axis. This results in surface and subsurface waterlogging. Consequently the subsurface salts along the prior channels come up to the surface through capillary action and convert the agricultural fields into saline lands. Interpretation of aerial photographs reveals that construction of a number of tanks in close proximity along a prior bed of the river Luni in the south-east of Jodhpur district have converted much area around Olwi, Rampuria and Ramassi into saline lands (Ghose et. al., 1975). Although a few non-saline cultivated fields still remain in the area, salinity is fast encroaching on them to make them unusable (Fig. 6.1). In Nagaur district the construction of an embankment across the prior course of the river Jojri at Kheduli has also resulted in the salinisation of the tank, spread of salinity to surrounding non-saline fields and consequent reduction in cultivable land in Ren-Kheduli-Kalru sector. About 86.4 sq.km. area has so far been turned saline. As the salinity is spreading laterally and down the valley, the farmers are changing their crops to salt-tolerant wheat. The irrigation of the fields, however, has not stopped in spite of the alarming nature of the problem.

At certain places the salinity has also developed due to the construction of canals across the course of a prior channel. For example, around Mori Sutran, Mori Manana and Kharabera Bhiotan the spread of salinity is due to the crossing of a prior channel by the Sumair Samand canal (Ghose et. al., 1975). The fresh water of the canal develops hydrostatic pressure on the subsurface saline water below it, which is then bulged upwards on both the sides of the canal and gradually spreads laterally.

Another example of man's activity leading to salinisation of agricultural lands is along the Luni river itself. Here the construction of a dam across the river at Pichiak has resulted in development of salinity, first around the reservoir. Then in less than 100 years' time the salt gradually leached down-stream up to Dunara — a distance of about 100 kms. Consequently all the wells along the Luni up to Dunara have now brackish to undrinkable water. Use of this saline water for irrigation at Kankani and Shikarpura have turned about 830 ha of agricultural fields into barren land (Singh et. al., 1978). At Dunara about 50 per cent of the wells are now saline and the rest yield sweet water. But the villagers apprehend that those remaining sweet-water wells will also turn saline in the near future.

Numerous other examples of salinity due to wrong irrigation practices exist in the Rajasthan desert, but the examples cited should be enough to make it understandable as to how much the use of a land for agriculture in this vulnerable ecosystem depends on the geomorphic principles and as to why not to tamper with the delicate balance of

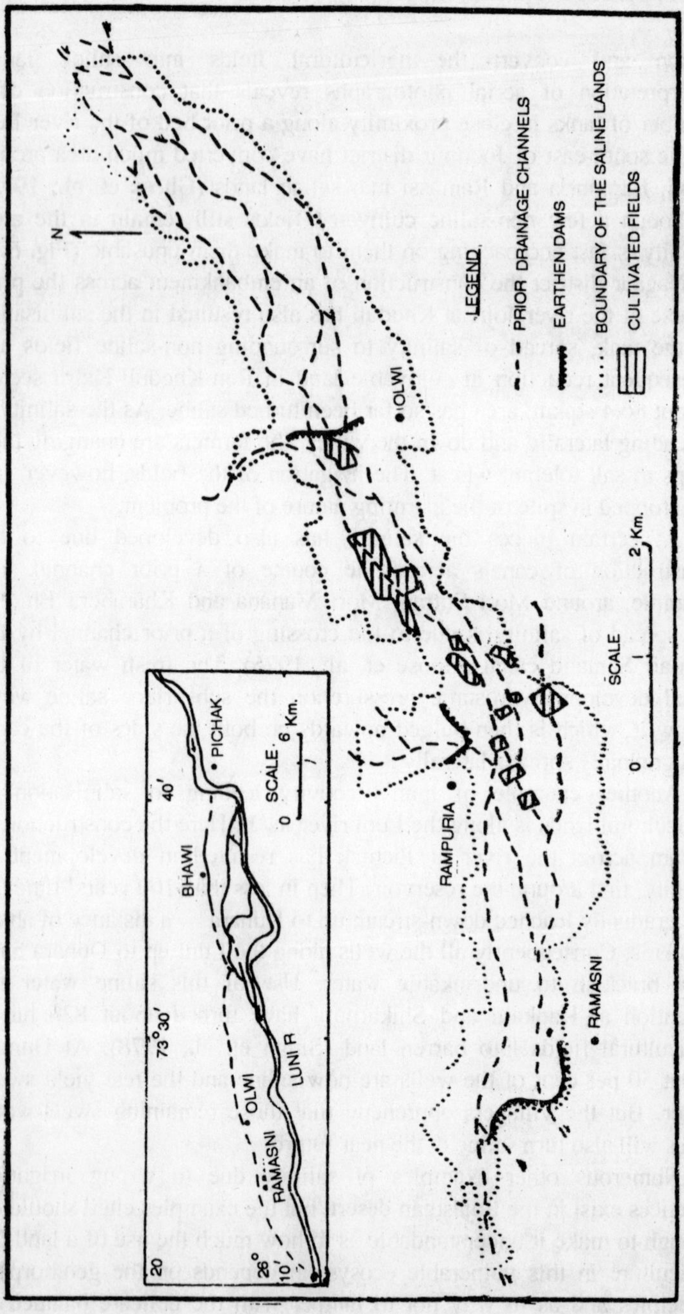

FIG. 6.1: Main-induced salinity along the prior drainage channels.

nature. Yet, more and more ambitious projects with little regard to the natural laws are coming up and threatening the precious cultivable fields with a doomed future. The glaring example is the Rajasthan Canal along the western border of India. According to a recent estimate the average rise in the ground water table in the areas through which the Rajasthan Canal passes is 0.55 metre per year (Mann & Chatterji, 1978). Surface and subsurface waterlogging are increasing due to the presence of impervious clayey gypseous bed underground. If the present rate of increase in the water table continues about 13,000 sq.km. area is going to have serious drainage and salinity problems within 50 to 100 years. The problem is more serious in Ganganagar district where the Gang Canal, constructed in 1927, has already threatened 2,930 sq.km. area and the rise in the water table in its command area is about 0.83 metre per year (Mann & Chatterji, 1978).

The Problem of Aeolian Activities

The other major factor limiting agriculture in the desert is the aeolian activity and its impact either in erosional or in depositional forms. However, the dynamics of wind is less understood for effective control of aeolian hazard. The sand dunes, the sandy hummocks or even the sandy plains react readily whenever they are disturbed by man, his animals and ploughing instruments. This is more so when the moisture content in the sand is less then needed for particle cohesion. It is also known that sediment particles with different densities and size move differently when confronted with wind of a particular velocity. For the quartz particle, which is the most common in any sandy area, the sizes most vulnerable to aeolian activity are roughly between 0.1 and 0.15 mm (Cooke and Doornkamp, 1964). The presence of silt, clay and decomposing organic matter in the soil, surface roughness in the form of clods, ridges and plant residues or a vegetation cover tend, on the other hand, to minimise the wind erosion. Not much work has been done in this country on the empirical relations between the velocity and other characteristics of wind and the nature and characteristics of sediments vulnerable to wind erosion, or how much finer particle and nutrient are lost from a particular area during a duststorm. The lack of such vital information does, of course, make it difficult to predict about the land which is more susceptible to aeolian hazard, both erosional and depositional, but from a critical examination of a field and its boundaries it can sometimes be pointed out if the field is subjected to net wind erosional hazard or to depositional hazard.

In the sandy plains, underlain by alluvium at shallow to moderate depth, the severe wind erosion piles up the moderate to coarse sand particles along the field boundaries and blows away the finer particles, resulting in ultimately the exposure of *Kankar* nodules from within the alluvium (Kar, 1978). The field then assumes a bowl shape. When this stage is reached the farmer becomes alarmed that his field is becoming unproductive.

On the other hand, the deposition of sand in a field is, to some extent, beneficial to a farmer, because the finer particles carry with them the much needed nutrients also. Moreover, the annual addition of sand to the otherwise sterile soil helps in better germination of the bajra seeds. Therefore, the farmers of a sandy plain prefer some amount of sand blown into their fields, rather than out of their fields. However, the continuous inflow of coarser particles to a field or the movement of a dune front due to crestal reactivation does harm a cultivator's field. On the sand dunes the indiscriminate lopping of trees and bushes, furrowing along the direction of the wind, or overgrazing by the domestic animals like sheep and goat loosen the structure and help in reactivation of sand. Examples of such menace are to be found in the western part of Nagaur tehsil of Nagaur district, in the south-western part of Jodhpur district and in parts of Barmer, Jaisalmer and Bikaner districts.

In the extreme north-west of Jaisalmer district the duny areas of the Bhuttewala-Ranahu-Tanot-Kishangarh sector is an ideal example of zone of deflation. Here the dunes are highly stabilised with profuse vegetation (mainly bushes) and take the form of swell and swale topography, the reason being that the deflation is minimum around the individual bushes whose roots act as a binding force, and is maximum in between the bushes where there is no such binding material. However, no man is responsible for this deflation. Where man's activity is responsible for the reactivation and movement of sand, suitable remedial measures like controlled grazing, introduction of suitable grasses and bushes to check the sand, minimum use of the dune slopes for cultivation etc., may be helpful. Creation of suitable wind brakes in the source area of sand movement does also help in restricting the movement.

Conclusion

The above discussion amply suggests that in the delicate ecosystem of the desert man is now becoming a significant geomorphic agent responsible for changes in the landform characteristics and also land utilisation. In many areas the changes brought about by man have benefited him, but in most other cases the cumulative effect of his activities has been more degradation of land — or in recent terminology the desertification, simply because of the lesser understanding of the process — response mechanism of the land. It has now been estimated from critical examination of the geomorphic indicators that while climatic boundary of the desert has shifted westward from its previous position (a trend towards de-desertification), man is aiding the desertification process in our country (Ghose et. al., 1977). However, the situation is not so alarming as has been described by Moore (1975): "The Thar Desert in India is also spreading, advancing at the rate of one-half mile a year along its whole perimeter. In 80 years an estimated 56,000 square miles, or an area equal to that of Wisconsin, has been turned into shifting sand" (pp. 59-60).

What is needed is better understanding of the geomorphic parameters of land, careful monitoring of changes taking place in the landforms under induced conditions and evolving of suitable technologies and landuse systems which will maintain the ecological balance.

REFERENCES

Cooke, R. U. and Doornkamp, J. C., 1974 : *Geomorphology in Environmental Management : An Introduction*, Oxford University Press, London, p. 413.

Ghose, B., 1964 : "Geomorphological aspects of the formation of salt basins in western Rajasthan", *Proc. Symp. Problems of Indian Arid Zone*, CAZRI, Jodhpur; pp. 79-83.

———, 1965 : "The Genesis of the Desert Plains in the Central Luni Basin of Western Rajasthan", *J. Indian Soc. Soil Science*, Vol. 13, pp. 123-26.

———, 1977 : "An Introduction to Desert Geomorphology", in Mann, H.S. (ed.), *Desert Ecosystem and its Improvement*, CAZRI, Jodhpur, pp. 65-72.

Ghose, B., Singh S. and Kar, Amal, 1975 : "Some Geomorphic Aspects of Salinity Hazard in Rajasthan Desert, India", *Workshop, Problems of the Deserts in India*, GSI, Jaipur.

———, 1977 : "Desertification around the Thar — a geomorphological interpretation", *Annals of Arid Zone*, Vol. 16, No. 3, pp. 290-301.

Ghose, B., Kar, Amal and Zahid Husain, 1978 : "Comparative role of the Aravalli and the Himalayan River Systems in the Fluvial Sedimentation of the Rajasthan Desert", Paper presented at the *Tenth International Congress of Anthropological and Ethnological Sciences*: Session on Tertiary and Quaternary Climatic and Environmental Changes, National Museum, New Delhi.

Kar, Amal, 1976 : "Man's intervention of the Desert Stream and its Consequences", *Proc. 63rd Indian Science Congress,* Part 3, Sect. 5, p. 20.

———, 1978 : "Geomorphic hazards : Their genesis, nature and problems of identification in the arid zones of India", *Proc. International Training Course on Integrated Natural Resources Survey*, CAZRI, Jodhpur. (Mimio).

Mann, H. S. and Chatterji, P. C., 1978 : "Effect of landuse changes on the ground water conditions in canal irrigated area of Western Rajasthan and their management", *Land and Water Management in the Indus Basin (India),* Vol. 1. Indian Ecological Society, Ludhiana, pp. 225-68.

Moore, Ruth, 1975 : *Man in the Environment*, Tata McGraw Hill, New Delhi, p. 155.

Ryabchikov, A., 1975 : *The Changing Face of the Earth*, Progress Publishers, Moscow, p. 205.

Singh, S., Ghose, B. and Kar, Amal, 1978 : "Geomorphic changes as evidence of palaeoclimate and desertification in Rajasthan desert, India (Luni Development Block : A Case Study)", *Man and Environment*, Vol. 2, pp. 1-13.

7

GEOMORPHOLOGY AND AGRICULTURE

ENAYAT AHMAD

GEOMORPHOLOGY IS so dominant a factor in the human environment that its influence on the pattern and destiny of agriculture is immense. This is true on macro-regional as well as micro-regional scales. Thus agriculture in mountains, plateaus and plains differs greatly. Similarly, scarps, pediments, valley flanks, flood plains, ridges and basins, are characterised by different types of agriculture. Micro-regional landforms like alluvial cones and fans, drumlines or eskers, monadnocks, levees, sand dunes and oases are marked by different agriculture.

As a matter of fact whether it be the case of a glacial or periglacial landscape or an arid and semi-arid topography or the land forms of a karst or coastal or volcanic region, the geomorphology sets the foundation and background of agriculture which can be adopted or can be feasible in a particular area. Now we may spell out these facts as follows.

Macro-Regional Geomorphology and Agriculture

Other things being equal, i.e. latitudinal location and stage of economic development including technological inputs, the geomorphology influencing the entire agricultural structure makes mountain agriculture different from plateau or plain agriculture. In mountains (i.e. highlands of steep converging slopes mostly more than 2,000 metres above sea level) the steepness of slopes is the greatest physical dominant. It affects the availability of soil moisture, depth, structure and texture of soil, the amount of solar radiation, the feasibility of terracing, field pattern, agricultural transport, possibilities of irrigation, the size and distribution of holdings, continuity of cultivation and the most important aspect of agriculture, i.e. the crops, their nature, yield and outturn.

Again on a plateau (i.e. a highland with steep edge at least on one side and a generally even or moderately undulating upper surface with small relief) has a markedly different physical setting because of the nature of slopes. The plateaus represent slope or topographic conditions which are intermediate between those of mountains and plains. Greater continuity of cultivation, development of zonal soils in contrast with the azonal soils of mountains, much easier and more frequent contact between farm and field, much easier communication, larger size of terraced field, and greater proportion of unterraced fields, as compared to mountain agriculture are some of the important differences resulting from the plateau geomorphology.

A plain (with extensive even surface and negligible relief) has such a geomorphology as ensures thick soil. Such topography rules out the deposition of very coarse sediments. It ensures greater sojourn of precipitated moisture and consequent greater soakage and greater ground water reservoir. Plain topography ensures continuity of cultivation. It facilitates the use of heavier agricultural machinery. Such a topography assures extensive irrigation because of low gradient. If we wish to consolidate holdings it is nowhere so more easily feasible as in a plain region. The working of fields, i.e. tillage, spading, hoeing, weeding, sowing and harvesting, is easier in the plains than in mountains and plateaus.

One of the important aspects of geomorphology is slope. Directly and indirectly slope influences agriculture in different manners. It determines terracing and its characteristics. It influences field pattern. It determines ground moisture and agricultural water supply. It influences the variety, yield and outturn of agricultural crops. It influences agricultural transport and practically all aspects of agriculture. Now let us elaborate these facts.

Slope and Terracing

Successful agriculture is possible only on a reasonably level ground. In sloping terrain such level land has been formed by terracing. Such terraces are found in all sloping (mountain or plateau or hilly) areas of the world since time immemorial. Originally when the farmer built a terraced field he must have cut the earth from a part of the slope and placed it immediately below so that part of the field was cut and part filled. If the slopes were gentle the terrace could be wider with the same amount of labour. For digging the same depth of slope the earth available would be much larger than in the case of steeper slope. Thus the terraced field would be large. This also means that for making the terraced field of the same area the terrace wall, i.e. the vertical bund at the lower end of the field, would be higher.

All these facts show how geomorphology influenced the operation, size and maintenance of terracing. "Terracing which is probably as old as agriculture itself mitigates the thinness of soil venner at any point and for the growth of normal agricultural crops which are generally shallow rooted, the available soil thickness is sufficient" (Ahmad, 1976).

Even a slope of 2° to 3° makes terracing inevitable particularly for the cultivation of transplanted paddy which requires level ground in which water should stay. Thus in Chotanagpur where the average angle of slope from the river bed to the adjoining upland in the village is about 2° to 3°. Terracing extends from the river bed to the crest upland. The gradient of the individual plots, however, is unrelated to the natural declination. It is on the other hand related to croppage. Thus on the lowlands where paddy predominates and rain has to be retained, the gradient of the individual fields is much less (approximately zero) than plain field growing millets or autumn rice where the gradient may up to 1° or 2°. This is true of the entire Peninsular upland.

Where the slope is steep as in the Himalaya terraced field may be as small as a "billiard table". In Peninsular upland where the slope is gentle on peneplains, the terraced fields may be much larger. "The greater the slope, the smaller the width of the field so that some fields are only 20′ (c 6 metres) wide and a third of a mile (c 530 metres) in length. This is rather an extreme case but plots one furlong (c 200 metres) in length and 80 feet (c 24 metres) in width are not uncommon. What is striking in this or other Chotanagpur villages is that contour bunding i.e. orienting terrace walls parallel to the local contour, is almost common and the bogey of contour bunding and contour ploughing will be superficially redundant for this scene" (Ahmad, 1976). This is probably true of the whole country and other parts of the world also.

Geomorphology and Field Pattern

Field pattern refers to the size, layout the alignment of cultivated fields. This is also greatly influenced by geomorphology. On level-plain surface the fields are rectangular but the rectangles may roughly be with equal sides. In a sloping ground the fields are terraced. The terrace walls have to be parallel to the local contour, otherwise the soil and water will not be conserved; they will be washed away. Thus the fields are generally unusually oblong in hilly terrains.

An aerial view of cultivation along the hill slopes or valleys in the Himalayas reflects the contour pattern of the country. Roughly speaking the size and width of plots varies indirectly with slope. On steep slopes while there may be difficulty in making the fields wide their length may be great. In such areas a series of long narrow fields form what may be termed as "stadium" pattern.

If the ground is dissected or rugged the cultivated field pattern may be patchy.

The rives may exert an elongating influence on the shape of the fields whose longer axis may become parallel to the river which is the source of silt and moisture.

On gentler slopes the fields may have a criss-cross pattern the plots being neither parallel nor rectangular, their boundaries running in a criss-cross fashion. The upper ends of valleys, gullies and ravines may be lobate and the field pattern is accordingly influenced.

Geomorphology and Ground Moisture and Agricultural Water Supply

Other things being equal, the amount of ground moisture varies indirectly with the degree of soil. Low gradient ensure slow discharge and consequent great soakage. The effect of steep slope is to waste away the flowing water by quick run-off. The agriculturist might have carried out experiments about ground moisture on different degrees of slopes and watched the growth and other conditions of crops, if not in India, at least in other developed countries. If not, they will have to do so in a world where the population is increasing and the relative pressure on land and agricultural resources is increasing without check. All technological possibilities and all physical controls will have to be carefully examined. The relation between slope and ground moisture is one such factor which must tell significantly on the growth, quality, yield and outturn of crops.

Agricultural water supply or irrigation is very much influenced by the form of land. Alluvial cones and fans are able to derive moisture from the snow fields and glaciers as in the glacial and periglacial zones of the Himalayas. Dissected terrain is a great impediment in the provision of irrigation to agriculture.

The provision of irrigation is the easiest in plain country. In dissected terrain there is no expansive area with gentle gradient to facilitate the construction of long irrigation canals. In plateaus short irrigation channels fed from storage reservoirs may provide irrigation but in mountain topography of high relief numerous topographic obstructions within a small area in the form of ridges and valleys form a strong limiting factor. In such areas the common means of irrigation as in the Himalayas is "gul" or "kul", a channel dug down the slope issuing from a point form a stream and passing through irrigable cultivation, following the local slope and joining the same stream further down. The gul or kul, therefore, is a diversion channel, the length of the channel depending on the permissibility of terrain.

A road can negotiate broken terrain. A railway too can negotiate moderate slopes or undulations but a canal can hardly do so. The former carry transport but the latter carries the liquid water, which can flow only on a continuously falling gradient. If water had to be raised at any point it can be done so at a high cost of pumping and necessary installations but from this point downwards again the principle of a continuously falling slope must be assured. Therefore, in high mountain and dissected, even the kuls or guls have to be short. Moreover, by its nature and source it has to keep close to a valley along the contour line.

The control of terrain on irrigation channels is decisive not only in the highly dissected mountain terrains. It is also abundantly visible even in moderately dissected terrains of Peninsular uplands. The canals are absent from the "steep scarp zones bounding the different peneplains and plateaus or the Western Ghats, or the hilly areas of the Aravallis or Nilgiris or the Eastern Ghats. In steeply sloping country or mountains, scarps or lower-ridge canals are generally out of the question for a number of reasons. There is hardly any extensive patch of level country of deep soil that needs water of the large volume involved in canals. The rugged terrain even for short lengths will have to be negotiated by numerous cuttings or tunnels and aqueducts which make the canal construction prohibitive. Such terrains are also associated with the hazards of landslides or deep erosion that will seriously damage canals" (Ahmad, 1971-72).

Soil — Foundation of Crops

Whether a soil is of zonal and drift type of residual and zonal with well-marked profile depends on geomorphology. The various physical and biotic processes which are essential in the evolution of soils or the development of their characteristics are influenced by geomorphology. The chemical and physical characteristics are also related to landforms and slopes. The texture and the size of soil grains are determined by geomorphology, its coarseness roughly varies directly with the degree of slope. Soil structure, i.e. the arrangement of the soil particles in different types of masses or bodies, e.g. nodular, laminated or massive, is also determined by geomorphology and slope. Other things being equal, humus content, soil alkalinity or acidity or other chemical properties are also related to slope and geomorphic location. Susceptibility to erosion or capability of conservation is also considerably influenced by the gradient of the ground. The thickness of soil cover is very much a function of geomorphology. Thus geomorphology by determining soil, which is the foundation of crops, influences agriculture to an immense extent in this respect also.

Geomorphology and Crops

By determining the gradient aspect, insolation and radiation, soakage and run-off of precipitation, soil characteristics and wind force and direction, geomorphology influences the type, growth, output and quality of agriculture crops.

As an example, saffron is grown in India only in Kashmir Valley where the local geomorphology of the Karewa plateaus above Pampur village is one of the decisive factors. "A particular aspect and sloping ground are required" (Sir George Watt). Cardamom luxuriates on undulating hilly ground with northern and western aspects in south India (Sir George Watt) at altitudes more than 600 metres. These facts bring out the control of geomorphology.

Among cereals maize is a crop which does not need much moisture in plains condition, this may be done by gently sloping the ground or by a large proportion of sand in the soil so that water is soon soaked downwards and does not accumulate at the roots. In hilly, mountainous and plateau areas there is no dearth of sloping ground as a result of which maize is the common crop of such geomorphic areas.

As a matter of fact all crops which require a limited amount of moisture are favoured by well-drained soils that characterise sloping hilly terrains. Thus what is provided by relatively arid climates in the plain topography is achieved by sloping terrain in the hilly areas.

Most of the Indian vegetables do not tolerate waterlogged conditions. That is why despite the poorer fertility of soils in the Himalayan and Peninsular upland regions most of the vegetables are grown in the mountains, plateaus and hills subject to the availability of other physical conditions particularly suitable climate and soil.

Such species which prefer sloping terrains as caraway (*zira*), *dalchini* (bark) whose leaves are called (*tej-pat*) and ginger benefit from geomorphology in the Lesser and Outer Himalayan zones.

Most fruit trees particularly temperate varieties shun waterlogged ground and benefit from the sloping terrains found in the Himalayas and elevated parts of the Peninsular uplands. Such temperate fruits as apple, pomegranate, walnut, pear, almond, peach, cherry and apricot prefer well-drained sloping ground.

It will be worthwhile to carry on experiments regarding the growth of different types of crops — cereals, fruits, vegetables, spices and other cash crops with varied geomorphology particularly slope. The physical conditions are so complex and subtle that it may not be possible to assess their relative importance without sustained experiments. Sometimes it may be seen that despite identical inputs the crops in two adjoining plots may be different as regards growth and yield. The harvest is a function of so many factors. Geomorphology particularly slope is one of them. The influence of slope on the crops can be experimented to determine the most ideal slope in a given situation.

In plain tracts the influence of geomorphology is seen in the difference between the cropping pattern of the riverside levee uplands on the one hand and the interfluvial lowlands behind levees on the other. The levees are marked by such crops as jute, maize, sugarcane, bhadai rice and oilseeds which prefer well-drained conditions. The lowlands or *chaurs* grow transplanted paddy and such rabi crops as benefit from the residual moisture of paddy lowlands.

The important sub-tropical crop, tea, needs flat or slightly sloping surface but waterlogging is fatal to it whence slope, even slight, is essential and a southerly and south-westerly (Sir George Watt) aspect is preferable. In such an aspect also geomorphology is implicit.

Geomorphology and Agricultural Transport

A dissected terrain makes movement of men and animals difficult. This is true of all areas — developed or developing. Even in the most advanced countries agricultural transport cannot be by helicopters. No doubt motor transport is common in developed countries but the present fuel crisis is a great check. In developing countries like India agricultural transport is by men and animals. Think of the conditions in the Himalaya or steeply sloping areas of Peninsular area of India like the Western Ghats or the various scarpments. Here it is difficult to move or to carry burden.

Agricultural transport includes the transport of agricultural implements, ploughing animals, fertilisers, seeds, etc. It also includes the harvested crops, crop straw or stalks or fodder. Agricultural products have also to be carried to the markets. In all these operations movement of articles may be performed by human beings either as headloads or on shoulders or, where the ground slope permits, on carts or beasts-of-burden or on levers or *banhgas*.

Those who have been to the mountainous or hilly areas must have realised the great hardship which is involved in such movements. This is one of the basic reasons for the dispersal of human settlements in rugged terrains. It would have been so difficult for the farmer to undertake agricultural transport from a central nucleated village to the fields in the outlying areas of the *mauza* that the only reasonable alternative was to build dwellings in different parts of the *mauza* nearer to cultivation.

We can realise how difficult it is to move up and down the slope especially with load. Gravity works against upward movement and causes strain and fatigue. In down-slope movement it hastens steps and works against sure-footedness. These difficulties vary directly with the angle of slope. Thus the more dissected the terrain and the steeper the slope the smaller the range of agricultural transport. In other words the more difficult the terrain the smaller is the distance between agricultural fields and habitations. This means the co-existence of dwelling and cultivation and the consequent dispersal of settlements.

The more dissected the terrain the lighter must be the means of transport and the quantity or size of agricultural transport. The rational animal, i.e. man, is probably the most reliable and successful means of agricultural transport in the most difficult terrains. Then are the small-statured light but sure-footed animals such as sheep, goats, mules, ponies and yaks.

REFERENCES

Ahmad, E., 1971-72 : "Geography of Irrigation in India", *Geographical Outlook*, Vol. VIII (Dept. of Geography, Ranchi University, Ranchi) pp. 9-28, 11.

———, 1977 : "Bhusur Village : A sample survey of landuse pattern and its physical setting in Chotanagpur", *Some Aspects of Indian Geography* (Central, Allahabad), p. 173.

Sir George Watt : *The Commercial Products of India*, Reprint edition (Today and Tomorrow, New Delhi), p. 429

Sir George Watt : Ibid., p. 514.

8

MORPHOLOGICAL PROCESSES, HUMAN INFLUENCES AND LAND DEGRADATION IN THE INDIAN DESERT

AMAL KAR

LIKE IN all other deserts of the world, the desert in India has also a fragile ecosystem, where varying degree of aridity and terrain constraints put severe limitation to agricultural and allied activities. Most of the terrain constraints could be viewed as the land degradation processes and associated natural hazards, which are again related to the present-day morphodynamic processes operating over the various landforms. Therefore, an assessment of the land degradation processes through correct identification of the present-day geomorphic processes, acceleration of the processes due to induced conditions, their manifestation through various erosional and depositional forms and mapping of the various degradational forms, alongwith their severity of occurrence, may provide clues to the problems involved.

The Processes During the Quaternary

The Exogenic Processes

From the southern part of Haryana and Punjab states in the north to the sea shore of Kachchh in the south, the hot arid belt of India (or the Indian desert) is dominated by the fluvial and the aeolian processes. Stratigraphic records from within the desert suggest that these two remained the major exogenic processes all through the Quaternary period also. In fact, early researchers suggested that aridity was responsible for disorganisation and disappearance of stream networks in and around the desert. However, the view that the Thar desert is of recent origin (Krishnan, 1952) or that the aridity is a 'post-glacial' (or Holocene) phenomenon (Wadia, 1960) is not valid now (Goudie, et al, 1973; Singh, et al, 1974; Ghose, et al, 1977; and Allchin, et al, 1978). The climate fluctuated all through the Quaternary, between drier and wetter phases, and resulted in changes in the relative dominance of fluvial and aeolian processes. Such changes were, again, a part of the global climatic changes. The glacial periods were the periods of maximum aridity and the inter-glacial phases experienced much less aridity (Fairbridge, 1968; and Sarntheim, 1978).

Evidence gathered so far from the satellite image interpretation and field studies, indicate that two major Himalayan streams, the proto-Saraswati and the proto-Drishadvati were in existence within the present desert in Rajasthan, atleast during the major wet phases of the late Pleistocene-early Holocene period when widespread alluviation took place (Ghose, et al, 1979; and Kar and Ghose, 1984). This was followed by a major dry phase that terminated around 10,000 years before present (BP). Most of the presently stable, high sand dunes in the region were formed during this period and drainage desiccation and channel changes took place. From 10,000 to 3,800 years BP the climate was wetter again. The fluvial processes gained prominence and there was a better stream organisation. The dry valleys of the Saraswati and the Drishadvati along the northern fringe of the desert were flowing streams during that period. A number of pre-Harappan (*c.* 4900-4700 BP) and Harappan (*c.* 4500-3900 BP) settlements evolved along these valleys, both in India and Pakistan. From 3800 years BP the climate became harsher again. The fluvial activities dwindled, the Saraswati disappeared at a place named Binasana (near Sirsa) in the Mahabharata, and most other streams were either buried under aeolian sand or were disorganised. Incidentally, the date of the Mahabharata (*c.* 3000 years BP) matches well with the date of this dry phase. As the aridity increased the aeolian processes gained prominence again, leading to the formation of low, mobile dunes and reactivation of many old and stable dunes, especially in the west. A number of inland sandy basins, which were created during the earlier dry phase and collected seasonal run-off since then, gradually turned into saline depressions, because of excessive evapotranspiration. A few of the major ones, like the one at Pachpadra or Didwana, had past links with flowing streams (Aggarwal, 1957; Ghose, 1964; and Kar, 1990a).

As a result of such process changes over time and space, a number of landforms have been sculpted to their present state. It also created many landforms which are neither compatible with the present-day processes, nor with the present zonation of processes. The major fluvially formed landform sequence is hills — rocky/gravelly pediments-buried pediments-flat older alluvial plains-younger alluvial plains-river beds. The activities of the aeolian processes over them created a new sct of landforms through the formation of sand sheets, sandy hummocks and sand dunes. Thus, parts of the flat buried pediments and flat older alluvial plains were transformed into sandy undulating buried pediments and sandy undulating older alluvial plains respectively, while the sand dunes and interdunal plains are the other major aeolian units. Table 8.1 summarises the broad framework of climatic change during the late Quaternary period and its effects on morphological processes and landforms.

TABLE 8.1: Geomorphic evidence of climatic change and desertification during the late Quarternary period in the Indian desert

Period in years BP*		Geomorphic processes	Climatic phases	Desertification/ De-desertification
O-Early centuries AD	Increasing fluvial activities, but increasing biotic stress on the land has resulted in salinity hazard along prior drainage channels and more deflation and deposition of sediments in the sandy plains and dune tracts	Biotic activities dominant	Slight amelioration of climate	Small scale amelioration of climate, but biotic desertification is present
Early centuries AD — 3800	Formation of low active dunes like barchans, shrub-coppice etc.; reactivation of the flanks and crests of the stabilised dunes. Inland basins turned into saline depressions	Aeolian	Dry	Climatic desertification
3,800 — 10,000	Major dunes stabilised; drainage channels revived; sandy inland basins filled with fresh water and fluvial sediments	Fluvial	Wet	Amelioration of climate leading to de-desertification
10,000 ?	Formation of major dunes; disintegration of drainage system by aeolian sand; creation of sandy inland basins	Aeolian	Dry	Climatic desertification
	Well integrated drainage systems; vast alluvial plains	Fluvial	Wet	No evidence of desertification

After Ghose, et al., (1977).

* BP means Before Present — After Singh, et al., 1974.

The Endogenic Processes

Apart from the climatic fluctuations the region also experienced several tectonic movements during the Quaternary period, leading mainly to the changes in base level and acceleration of the fluvial processes. In the Rajasthan part of the desert a set of two terraces along many streams of the Luni-Jawai plains and anomalous channel behaviour across the lineaments bear testimony to atleast two episodes of movement during the late Quaternary period (Kar, 1988a; 1989a). Evidence of neotectonism has also been found in the Mendha river basin along the eastern margin of the desert (Dassarma, 1986), while archaeological excavations at Kalibangan in the dry valley of the Saraswati revealed evidence of faulting in the alluvial stratigraphy, encompassing the pre-Harappan settlement, but not the succeeding Harappan settlement (Lal, 1979).

In the south the Kachchh mainland, the Great Rann of Kachchh and the Little Rann are experiencing earth movement throughout the Quarternary period. Biswas (1971, 1974) identified a series of E-W faults and a few N-S ones, which control these movements. During each movement since the Tertiary the E-W fault blocks are getting more uplifted along their northern margin and tilted southward (Biswas, 1971), leading to the formation of a series of cuestas across the Kachchh mainland, with north-facing escarpment and southward dip slope. The three major bets (islands) within the Great Rann, namely the Bela, Khadir and Pachchham bets, have also steep north-facing escarpments along the hill ranges in their northern extremity and gently dipping southward slope. Consequently, all over Kachchh the south-flowing streams are vigorously cutting headwards through the Quaternary alluvium, the Pliocene Kankawati and other uplands and the Deccan Trap hills beyond, leading ultimately to widespread stream capture within the central hilly tract on the Mesozoic formation. The north-flowing streams are adjusting their courses to the movements through terrace formation and channel incision (Kar, 1989a). Once out of the hilly tract the north-flowing streams tend to bifurcate into distributaries and create alluvial fans with bird's foot arrangement. Among the N-S structural trends, movement along the hinge zone from Mandvi-Mundra in the south to Pachcham island and beyond in the north is gradually tilting down the western part of mainland during each episode of movement, while uplifting its eastern part (Biswas, 1971). Consequently, the coastline from Jakhau westward is a drowned prograded coast, while the coastline to the east of Mundra is an emergent one, with wide tidal mud flat. The hinge zone of Mandvi-Mundra has numerous recurved spits and bars, as well as a major cuspate foreland at Mundra. Although the spits are forming here in response to a strong east-flowing longshore current, their clustering along the hinge zone suggests some control of the neotectonic movements in furtherance of the cause. Indeed, the backshore zone here is dominated by two distinct raised beaches, the higher one topped by the coastal foredunes.

The Great Rann and the Little Rann are the relics of a former sea, that receded southward gradually during the Quaternary period, mainly in response to the upheavals along several structural lineaments across them. The lineaments could well be marked through satellite image interpretation. Many of the smaller *bets* within the Ranns (e.g. Nara, Dhan, Boria, etc.) are raised mud flats, sometimes with coarser aeolian sand cover. The *Banni* grassland between the mainland and Pachchham island is also a raised mud flat with southward tilt. Archaeological excavation at Dholavira in Khadir island with the Great Rann is now revealing the ruins of a Harappan and pre-Harappan township, built, perhaps, by a sea-faring people. The earthquake of AD 1819 in the region activated the E-W lineaments through the Great Rann and the Indus-Saraswati delta, threw up an earthen mound of 3 to 6 m height, subsequently known as the Allah Bund (Burnes, 1835), beheaded the dwindling Saraswati (the Nara in Pakistan) and its many smaller distributaries and partially drowned the mouth of the river, known as Sir Creek (Kar, 1989 e). Consequently Lakhpat, which was a major part at the Creek head, became a deserted place.

The Present-Day Processes

As noted earlier, the fluvial and aeolian processes are still the major ones in the Indian desert. However, the relative spatial dominance of the two is guided by the present distribution of rainfall. The gradual change of climate from a relatively wetter east along the Aravali ranges to a far drier, extremely arid situation along the Indo-Pakistan border in the west, has enormously influenced the efficacy of the processes. The fluvial erosion is dominant along the eastern margin, but is totally replaced by the aeolian erosion/deposition along the western margin, while in the central part there are competing and complementing fluvial and aeolian processes.

The Fluvial Processes

Generally the major climate-related processes in an area undergoing changes from an arid to semi-arid condition, or vice versa, are expected to react in the following manner over time:

Climatic phase	= Wet → Transition →	Dry
Morphological activity	= fluvial Fluvial	aeolian
	erosion → deposition →	accretion

Conversely,

Climatic phase	= Dry → Transition →	Wet
Morphological activity	= aeolian fluvial erosion & deposition → lacusrine →	Sand stabilisation formation

The eastern wetter margin of the desert, experiencing former dry climatic phase and, hence, having stable aeolian forms, is acted upon vigorously by the fluvial processes, creating a finer drainage net of rills and gullies and producing increased sediment load from its sandy terrain. The depth of dissection is generally more in the sandy undulating terrain and on the major obstacle dunes along the Aravalli ranges, because of the stability of the surface and greater slope. In comparison the adjacent non-sandy terrain has a much coarser drainage net and hence, lesser erosional hazard. However, the efficiency of the fluvial processes decreases sharply after a certain distance away from the Aravallis, as most of the streams originating from the Aravallis either die out after flowing for a certain distance over the desert plains, or are integrated to provide a much coarser drainage texture.

Lithological variation and structural properties do also help in increased dissection in the east. For example, the pre-Aravalli granite-gneiss, which occurs discontinuously along the Aravalli foothills, has a low undulating terrain dissected by numerous rills and gullies (Kar, 1983). Neotectonism has also resulted in an increase in the drainage density. For example, in the Rajasthan part of the desert movement along one ENE-WSW lineament through Sirori, Revdar, Mandar and Virol has rejuvenated the south-flowing tributaries of the Sipu from the Jaswantpura hills and uplands, with consequent stream incision and development of an intricate network of rills and gullies on the obstacle sand dunes and whalebacks in the area (Kar, 1988a). As has already been noted, similar rill and gully erosions controlled by neotectonism, are degrading the terrain in the Kachchh mainland also and increasing the problem of soil erosion. In the coastal alluvial plain of Kachchh a series of broad Guaternary arches are developing as a result of neotectonism. There is a gradual thinning of alluvium towards the top of those arches and Plio-Pleistocene rocks show appearance. Associated with this change is the sudden increase in sheetwash and gully activities towards the top, as the major channels try to cut through the upliftment (Kar, 1989a).

Compared to the rill and gully erosion, the manifestation of sheet erosion is not so prominent, nor is there enough information on the subject. However, in many parts of the flat older alluvial plains, especially between Nagaur, Jodhpur, Pali and Merta (roughly between the 350 mm and 250 mm isohyets), the author noted areas of exposed lime concretionary nodules which generally form in the 'B' horizon of the older alluvium in the desert, because of insufficient leaching of calcium from the profile. This lime concretionary layer, which is also known as the *Kankar* pan in this part of the world, occurs at a variable depth within the older alluvium and as a hard strata. This often is the limit for agricultural operation. The exposure of this layer indicates the magnitude of sheet erosion in the locality and consequent loss of alluvial cover (Kar, 1989b).

Apart from the above types of fluvial erosion the spasmodic floods in the desert cause stream bank erosion along the Luni and its major tributaries in the eastern part of the desert. The dominantly sandy banks of the streams usually lead to channel braiding, although erosion of the concave banks in the meandering section is also dominant. The enormous sediment load, generated during the short spell of run-off, does also get deposited within the channel bed and the younger alluvial plains, leading to channel shifting and occasional infertility of the younger alluvial plains. The 1979 flood (a 100 year phenomenon) in the Jojri river near Jodhpur buried the fertile younger alluvial plains along the river with 1 to 2m deep inert coarse sandy deposits of the river bed (Dhir, et al, 1982). A number of dug wells along the river were also completely filled up, further accentuating the problem.

The Aeolian Processes

The movement of sand by wind, or the aeolian activity, is a major problem in the desert. Although erosion and deposition are the two inter-related aspects of the aeolian processes, it is often very difficult to classify areas into zones of exclusive erosion and deposition. This is because of the complex relationship between the wind velocity, air flow pattern, terrain characteristics, particle size and vegetative cover. For example, a short period of measurement in a field may register aeolian erosion during a high velocity regime, but the same field may again register sand deposition during low flow. According to Woodruff and Siddoway (1965) wind erosion is a function of several factors :

$$E = f(I.K.C.L.V)$$

where *E* is erosion in tone/acre/year,
I is soil erodibility index,
K is soil ridge-roughness factor,
C is wind erosion climatic factor,
L is unsheltered median travel distance of wind across a field, and
V is equivalent vegetation cover.

However, the complexities involved in the factors make the solutions difficult. Skidmore (1988) provided an easier guide for assessing the wind erodibility of bare soils, based on the proportion of dry soil aggregates of more than 0.84 mm size.

The manifestation of aeolian processes are mainly in the form of sand dunes, smaller sandy hummocks or sand streaks and sand sheets, all of which are depositional forms. Since the major stable and high sand dunes were formed during an earlier dry phase and the low, mobile dunes like barchans, barchanoids, sandy hummocks, other longitudinal and transverse streaks and sand sheeting are taking shape during the present phase, alongwith the reactivation of many stable dunes of the earlier dry phase, mapping of these forms provide a better clue to the dominance of the aeolian processes. It was found that the aeolian activities are dominant mainly to the west of 250mm isohyet and partly between 350mm and 250mm isohyets. High megabarchanoids (20 to 45m high) and sand dunes of considerable height are now forming to the west of 150mm isohyet (Kar, 1987a. 1990b). Isolated barchans of 2 to 5m average height are forming spontaneously in the less than 200mm isohyet and move NE ward at the rate of 30 to 50m per year (Kar, 1989c).

Using a climatic wind erosion index, originally proposed by Chepil et al, (1962) and modified by Yaalon and Ganor (1966), Kar (1989c) demonstrated the nature of erosivity of the wind along a transect from Jaipur in the east to Jaisalmer in the west (Table 8.2).

TABLE 8.2: Climatic wind erosion indices for some stations of the Indian desert and its fringe

Station	Climatic wind erosion index (%)		
	(a)	(b)	category
Jaisalmer	32.59	73.78	High
Jodhpur	3.71	6.81	Very low
Nagaur	2.30	4.58	Very low
Jaipur	1.54	2.44	Very low

(a) is based on mean annual wind velocity;
(b) is based on mean wind velocity of summer months when SW wind blows (March-September).

The results are in conformity with the conclusion drawn earlier on the basis of observation of forms.

Evidence of net aeolian erosion is difficult to come across in the sand covered areas, due to a general lack of reliable marker. However, exposed plant roots, exposed calcretes in sandy alluvial plains, or bowl shape of the cultivated field with bounding, high fence line ridges provide clues to net aeolian erosion. A knowledge of the type of air flow in the dune-covered areas and around other obstructions can also help in determining the zones of erosion and deposition. Although a detailed discussion on this aspect is beyond the scope of this article, it has been possible to suggest why certain traffic routes in the lee shelter to high hills in the dominantly sandy terrain remain relatively free of sand accumulation, while those further away from the hill, but still in the lee of the hills should attract sand accumulation (Kar, 1989c; Kar & Singh, 1989 and Kar & Joshi, 1989). One of the major reasons for formation of many of the saline lakes in the eastern part of the desert was a slow deflation at the lee of hills, although stream trapping by advancing longitudinal dunes along the margins of those hills was another responsible factor (Kar, 1990a). Similar aeolian process-form relationship for the genesis of saline lakes was also noticed in Barmer district in the west. The same principle of air flow behind the high obstructions also results in deflational plains (and even hollows) at the lee of major parabolic and transverse dune chains in the central and western parts of the desert. Deeply scoured and deflated interdunes in the extremely arid south western part of Jaisalmer district, which collect occasional rain water and then become pools of brine (locally called as *Dhand*), are also the results of wind erosion (Kar, 1990b).

The Other Processes

Apart from the above processes, a slow weathering of rocks and minerals is continuing throughout the desert and producing fragments ready for other processes to act upon. The rocky uplands and pavements of Pokaran-Jaisalmer-Ramgarh tract in the less than 200 mm rainfall zone are aplenty with numerous mechanical weathering forms like ventifacts and splitted boulders and chemical weathering forms like box-work and honeycombs, as well as desert vernish (Kar, 1989d). The above chemical weathering forms and many others like tafonis and pedestal rocks are numerous in other parts of the desert also, especially in the granite and sandstone terrains. However, most of these have no direct relationship with hazards. The slow subsurface weathering of the Jalor granite and pre-Aravalli granite-gneiss, mainly in the upper Luni basin, is on the other hand, resulting in development of salinity-alkalinity in the overlying older alluvial plains (Kar and Shankarnarayan, 1984). The shore zone of Kachchh continues to be influenced by the fluvio-marine processes, although episodic neotectonic activities do also take place. The Mandvi-Mundra-Anjar coastal belt regularly experiences earthquake shocks, a few like that of 1950 being of moderate to high intensity. We have also mentioned about the 1819 earthquake and its effects. A strong east-flowing longshore current is producing spits and cuspate foreland along the coast.

Human Influences, Process Acceleration and Land Degradation

Superimposed on the scenario of activities of natural processes in the fragile environment of the Indian desert are the myriads of human activities. Since the terrain is inherently vulnerable and the pressure on it is very high, the processes get accelerated in many parts and result in land degradation at a faster rate and wider scale.

Accelerated Fluvial Erosional Hazard

Along the wetter margin of the desert the Aravalli hill slopes are being regularly denuded of natural vegetation cover. Even the notified 'reserved' and 'protected' forest areas are being regularly cleared by the local people, because of the pressure on existing agricultural lands and scarcity of other resources. Consequently, the regolith cover on the hill slopes is being gradually washed down through sheet, rill and gully erosion, as the run-off is increasing in the absence of soil binding root-systems. Unfortunately the deforestation continued through the centuries and decades, without an adequate replantation programme and resulted in the present state of highly degraded and barren slopes without soil cover.

Many isolated hills, flanking the Aravallis in the eastern part of the desert, are also facing similar threats. These hills with lesser regolith cover and a more vulnerable ecology, have now a completely barren look with moderate to severe water erosion and almost total loss of regolith cover. The hills further west do not have appreciable regolith cover and are naturally barren, forming rocky wastelands. The pediments are buried pediments, flanking the hills, are also affected by sheet, rill and gully erosion and, hence, require adequate soil and water conservation measures (Kar, 1987b).

Accelerated Aeolian Hazard

The increasing biotic pressure on the land in the east is also reactivating the stable sandy landforms. Most of the sand dunes in more than 300mm isohyet zone were formed during an earlier dry climatic phase. Along the foothills of the Aravallis and farther east these dunes could well be termed as fossil dunes, because of their stability, vegetation cover, profile formation in them and dissection by rills and gullies. Yet the land is vulnerable due to the sandy nature of the terrain and there is a limit to the capacity of the sandy forms to maintain their balance with the exploitative enterprises. Although that limit is seldom exceeded in the wetter east, because of the resilliance of the terrain and the natural vegetative growth, even under the present trend of exploitative enterprises, wanton destruction of vegetation on the stable dunes by the local people and trampling of the dunes by their domestic animals sometimes cause serious problems of sand reactivation and associated land degradation (Kar, 1987b; and Kar and Shankarnarayan, 1984). This is precisely why the Pushkar-Budha-Pushkar lake region near Ajmer is now facing accelerated sand movement from the erstwhile stable dunes and sandy plains, formation

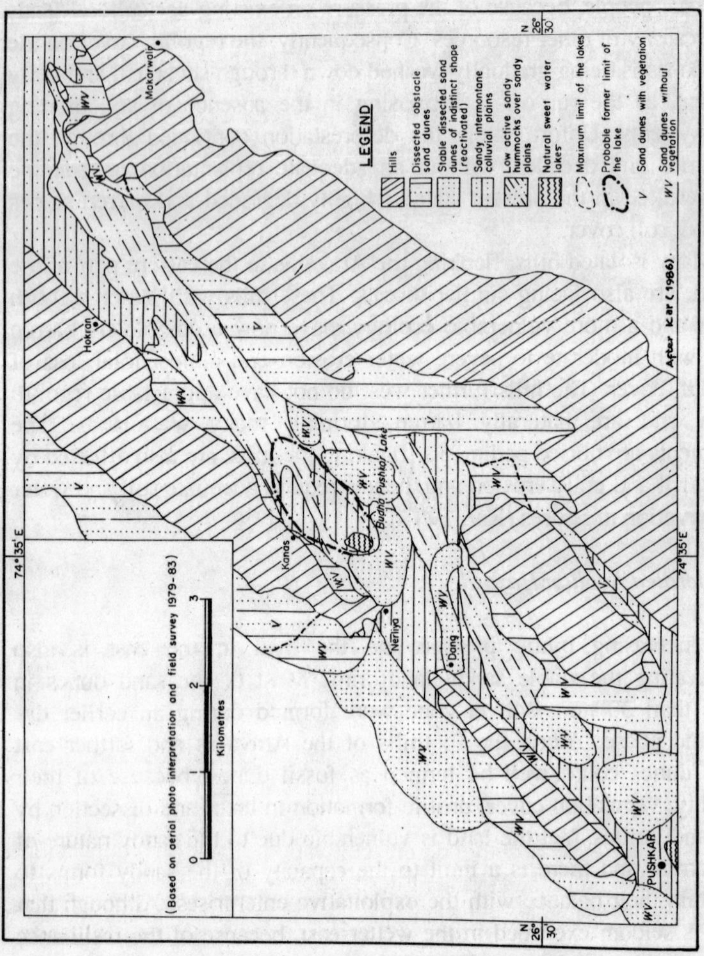

FIG. 8.1· Detailed morphological map of Pushkar-Budha-Pushkar lake region (India).

of new, mobile, sandy bed forms and shrinkage and deterioration of the natural lake environment, inspite of the area being situated in the 500 mm rainfall zone (Kar, 1986; Fig. 8.1). Such destructive trends could still be reversed through well planned programmes of afforestation, soil and water conservation and people's participation. Similar degradation of the aeolian landforms has taken a serious proportion in Rajgarh-Churu-Sikar tract, inspite of its better rainfall status.

In the central part of the desert, investigation in Luni development block of Jodhpur district (Singh, et al, 1978) indicated that leaving the agricultural fields barren in the sandy alluvial plains during the summer months, or deep ploughing through tractor, especially in the direction of wind, encourage moderate to severe wind erosion from the fields. Consequently, the finer particles are blown away from the fields and medium to coarse particles are trapped along the fences around the fields, leading ultimately to the exposure of the *Kankar* pan as an extreme event.

The field assumes a bowl shape and becomes much less productive (Kar, 1981). The farmers generally put up high fences around their fields in the sandy terrain to trap as much sand as possible, which blows out from their fields during the sandstorms (*aandhi*), and retrieve it after the sandstorm. Creation of such barriers and compartments, however, favour box—like pattern of massive wind erosional zones within the fields and depositional zones along the fences, which create high sand ridges in a network pattern. Such man-induced aeolian bedform development is especially dominant in the sandy plains of Lohawat-Agolai-Balotra-Luni-Jodhpur area in the central part and Nagaur-Kuchaman-Sikar-Ratangarh-Sujangarh area in the east.

Elsewhere within the desert cultivation along the dune slopes, overgrazing and other forms of destruction of natural vegetation on the dunes are reactivating the stable dunes and encouraging the formation of new mobile dunes. Such hazards are more conspicuous around the settlements. In the Indira Gandhi Canal command area, lying mostly in the less than 200mm rainfall zone, the stable dune ecosystem has been disturbed through canal digging operation, clearance of vegetation and land levelling for cultivation. Although the canal promises to bring agricultural prosperity to the region, immense problem has presently been created by the reactivation of erstwhile stable dunes, threatening roads, canal networks and settlements (Kar, 1989 b,d). A massive programme has now been taken up to check the menace along the canal. However, the land levelling and devegetation of the dunes by the individual farmers in the hope of assured supply of water that may not reach the devegetated and levelled land, are creating problems of new barchan formation in the erstwhile flat interdunes. This is a problem which could have been easily avoided through better planning. Away from the canal-related activities the sand dunes have remained much less affected.

Studies carried out by CAZRI have yielded some quantified information on the accelerated aeolian erosion due to cultural practices. For example, during the sandstorm of June-July, 1985 (ranging for 17 to 25 days) in the western part of the desert, the deep ploughed sandy plains, using tractors, lost 2630 to 3160 tonnes of soil per ha, whereas the lands lying fallow for a year or more, and having 2 to 4 per cent vegetation cover, lost 207 to 283 tonnes per ha. The pasture which could maintain 8 to 12 per cent vegetal cover, remained almost protected (Dhir, 1989). At Chandan in Jaisalmer district the clearing of vegetal cover from a plot with thick sandy soil led to the loss of 15 to 18cm thickness of top soil or 3128 to 3754 tonnes/ha during that sandstorm (Kar, 1989c).

Accelerated Salinity-alkalinity Hazard

Apart from the above accelerated fluvial and aeolian hazards, the other most damaging problem within the desert is the salinity-alkalinity build up in the alluvial plains, aggravated in most cases by anthropogenic factors. For example, in the medium to heavy textured alluvial plains of the upper Luni basin the five irrigation reservoirs at Pichiyak, Nayagaon, Sardar Samand, Kharda and Hemawas, are either collecting saline water, or are favouring salinity in the reservoirs because of salt-releasing formation underneath. The maximum damage to the land has been done by the Chaupra Raj Sagar at Nayagaon where the Chaupra canal brings the surface flow of the Raipur Luni through a shallow granite terrain topped by heavy textured alluvium. The saline flood water of the Luni is also being collected at this reservoir through an abandoned channel from near Parasla. Irrigation with this water in the medium to heavy textured alluvium has rendered many fields saline and unsuitable for cultivation (Kar and Shankarnarayan, 1984; Kar, 1987b). The Sardar Samand tank, constructed at the confluence of the Sukri and the Guhiya, is also gradually becoming saline and has affected its command area. The major responsible factor is the Guhiya river which carries very high salt load compared to the Sukri which, originating from the Aravallis, carries sweet water (Shankarnarayan and Kar, 1983). The Kharda and the Hemawas tanks have also been constructed in salinity-prone Jalore granite terrain and hence their command areas are also gradually becoming saline. Apparently all these tanks were constructed without proper assessment of all the relevant environmental parameters.

Problem of salinity-alkalinity is also increasing due to the construction of earthen embankments and canals across the former buried drainage channels, having subsurface brine along their courses (Kar, 1981; Ghose, et al, 1982). The problem has been noticed mainly in the Luni basin, especially along the former courses of the Luni. The construction of the embankments across those courses to conserve seasonal run-off along the slight topographic lows of the valleys results in surface and subsurface waterlogging. Subsequently the subsurface brine comes up to the surface through capillary action and converts the agricultural fields into saline wastes. For example, the problem has been faced along the Agariya Nala and other former courses of the Luni, around Rampuria, Ramasni, Olwi and Nimbli in Jodhpur district. In Nagaur district about 87 sq. km. area along the dry valley of the Jojri around Ren, Kheduli and Kalru has already been salinised due to the construction of embankments across the dry valley at Kheduli (Kar, 1981). Elsewhere, irrigation with ground water having high residual sodium carbonate has threatened many fields with sodicity problem, while dumping of industrial effluences in the rivers (e.g. the Luni at Balotra, the Bandi at Pali and the Jojri near Jodhpur) is increasing the toxicity level of the water, polluting ground water along those streams and spreading salinity problem in the fringing younger alluvial plains.

The major area of salinity-alkalinity build-up within the desert is, however, in the command area of the Indira Gandhi Canal in the west. Most of the command area, although sandy in the surficial layers, has a gypsiferous hard pan and/or ill-drained clay bed within the alluvial plain of the now extinct Saraswati (including the dry valley of the Ghaggar) that formerly used to drain the area. Profuse irrigation in such terrain, without proper arrangement for soil drainage, is not only creating the problem of waterlogging, but also spreading the salinity-alkalinity problem. The problem is more serious in Ganganagar district where the Gang Canal, constructed in 1927, has already affected more than 2900 sq.km. area and the rise in water table is between 0.67 and 0.83 m/year (Mann and Chatterji, 1978). Seepage from the canal and spreading the water in the alluvial plain underlain by gypsiferous bed at shallow depth are also causing drainage impedence and salinity (e.g. at Dabhli, Sansardesar, Bithu). Since groundwater movement through such a medium is unpredictable, the problem is difficult to contain once it sets in, especially when confronted with continuous discharge from the canal.

The problem was particularly accentuated when eighteen interdune plains of Suratgarh-Baropal-Kanaur area in Ganganagar district were linked to the Ghaggar valley for avoiding the menace of flood in the area and in the hope of absorbing the excess water in the supposedly deep sand dune and interdune landscape (Plate 8.1). The interdunes, however, have hard gypsiferous bed at 2 to 4m depth, which created drainage impedence, turned the fertile interdunes into vast lakes with weed infestation and gradually seeped out in the fringe areas, flooding and salinising the fields and settlements in Baropal, Manak-Theri, Kanaur and other nearby villages. Although the practice has very recently been stopped, it will take a long time to reclaim the land.

A third kind of problem is coming up near Madasar in Jaisalmer district (Stage II of the command area), where the main canal crosses a former dry valley of a stream that had its origin in the saline *rann* of Bap and used to flow northwestward (Plate 8.2). Because of its connection with the saline *rann* the stream's course, although presently filled with aeolian sand, is full of subsurface brine. Now, seepage from the canal near Madasar is gradually bringing the salts to the surface and spreading it northwards within the command area (Kar, in press).

At the extremes of their scale all the above processes are creating wastelands (Singh and Kar, 1988) and considerable environmental problems, leading to desertification (Fig. 8.2).

Conclusion

The Need for Understanding the Process-form Interactions

The above account of the process-form interactions in the Indian desert brings to the fore the fragility of its environment and the role of man in accentuating the problems. However, the scenario sketched above is not intended to provide a pessimistic outlook of either the present or the future. Although human follies have resulted in or accentuated many of the problems, human endeavours are also helping in assuaging the environment of man. For example, numerous tube wells dug by the

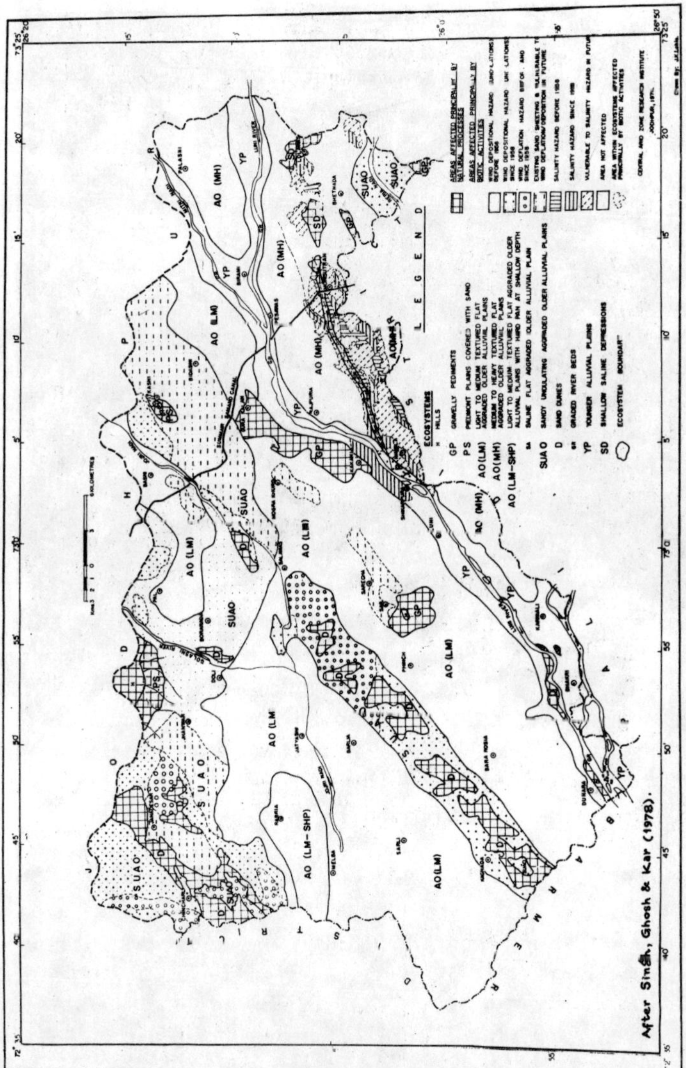

After Singh, Ghosh & Kar (1978).

FIG. 8.2: Areas affected by desertification processes in Luni development block.

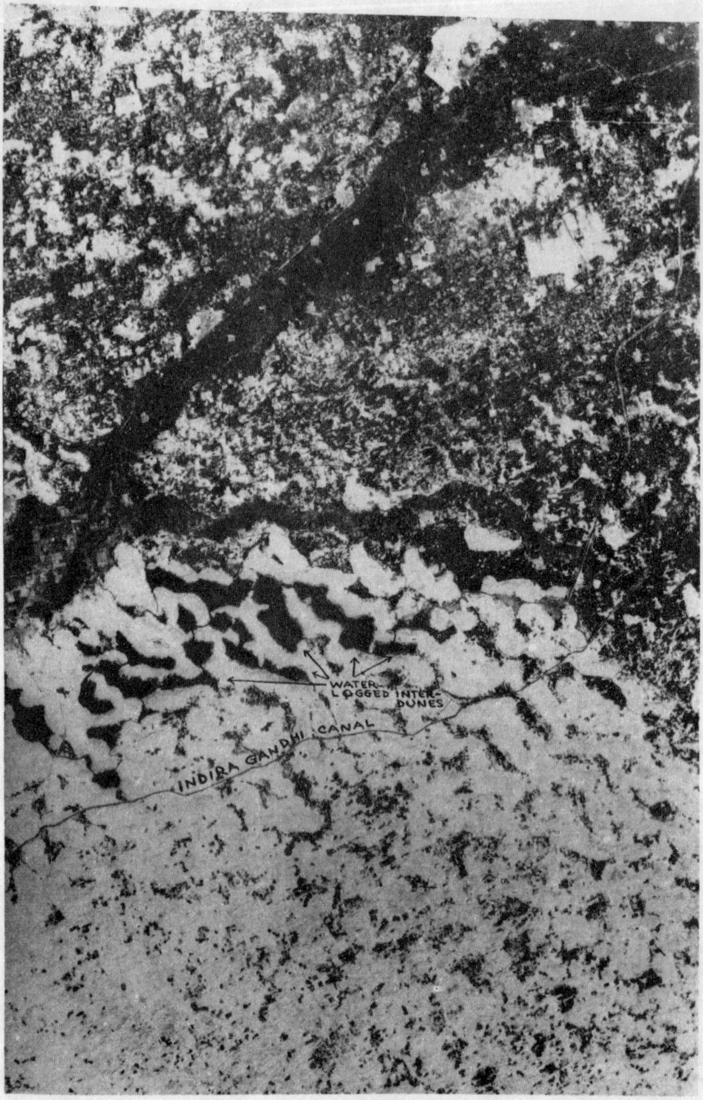

PLATE 8.1: Landsat TM subscene of Suratgarh area (band 3), showing waterlogged interdune plains due to diversion of canal water.

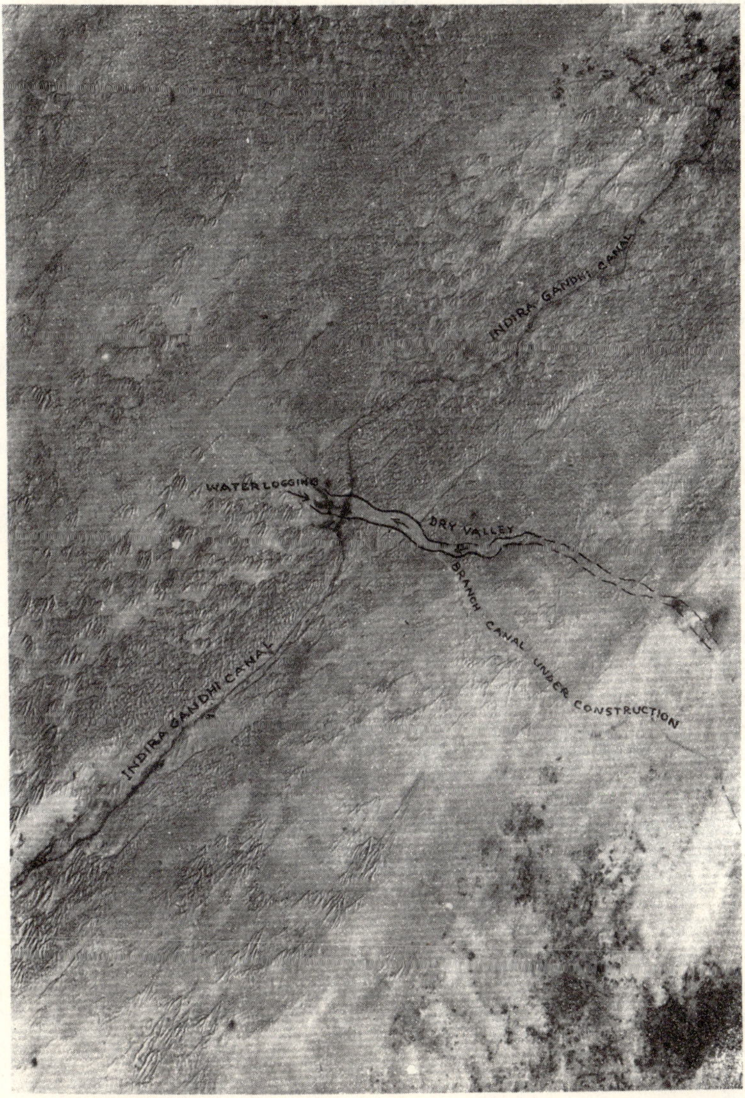

PLATE 8.2: Landsat TM subscene of Madasar area (Band 3), showing seepage from canal at the point of intersection with a dry valley and resultant spread of water and salinity to the north.

government agencies now meet the requirement of drinking water and irrigation in many areas. The people are learning new techniques of agriculture to improve upon their past performances, while the Indira Gandhi Canal network promises to revolutionise the land utilisation pattern in the extremely arid part of the desert. Yet, because of the fragility of the terrain it is necessary to understand properly the process-form interactions and the environmental impact of the developmental inputs before any plan is executed. The damages caused by some of the irrigation schemes should be a pointer in this direction. Moreover, the ill effects of indiscriminate felling of trees and removal of bush/grass cover from the sandy terrain, or profuse irrigation even without provision for soil drainage in the vulnerable tracts, need to be understood. The analysis also reveals that blaming man for every case of degradation is unjustified. Some of these take place spontaneously under natural settings.

Classification of Hazards for Mapping

A two-tier, 17 unit classification of the different erosional, depositional and salinity-alkalinity hazards within the desert has been prepared, wherein seven major types of hazard, namely sheet, rill, gully and stream bank erosional, wind erosional and depositional and salinity-alkalinity hazards, have been identified. The types have again been suitably graded according to their degree of severity. Combining both the factors (type and degree) we can have a number of mappable hazard classes (Kar, 1988b; 1989b). Some probable classes are as follows:

I. *Water erosional hazards:* (1) Slight to moderate sheet to rill erosion; (2) slight to moderate rill to gully erosion; (3) moderate to severe sheet to rill erosion; (4) moderate to severe rill to gully erosion; (5) severe to very severe sheet to rill erosion; (6) severe to very severe rill to gully erosion; (7) slight to moderate stream bank erosion; (8) moderate to severe stream bank erosion; (9) severe to very severe stream bank erosion.

II. *Aeolian hazards:* (10) slight to moderate wind erosion; (11) moderate to severe wind erosion; (12) slight to moderate wind deposition; (13) moderate to severe wind deposition; (14) severe to very severe wind deposition.

III. *Salinity-alkalinity hazards:* (15) slight to moderate salinity-alkalinity; (16) moderate to severe salinity-alkalinity; (17)

severe to very severe salinity-alkalinity.

Several other classes could also be worked out, as well as combinations of classes, depending upon the scale of mapping and complexity of the problem. In other words, the classification is flexible enough to accommodate new units.

Utility of Remote Sensing in Mapping the Present Status and Monitoring the Changes

Maximum mappable information for this type of study is obtained from remote sensing. Although large-scale, conventional panchromatic aerial photographs are most helpful in mapping the units, the time taken for interpreting the huge number of photo pairs for district-level or even block-level mapping and correct positioning of the interpreted data on the base maps is much more than what could now be achieved through satellite remote sensing, because of the scale-corrected synoptic view of the terrain provided by the latter and their improved resolution. The restricted nature of the aerial photographs is also an obstacle which one does not face for obtaining the satellite data products. Experiences indicate that the False Colour Composites (FCC) of bands 2+3+4 (spectral wavelengths in micrometer: 0.52-0.60, 0.63-0.69 and 0.76-0.90 respectively) of the Landsat TM (or Thematic Mapper Sensor) with 30m resolution are very useful for these studies, especially when the data products are available as prints at 1:250,000 or 1:50,000 scale or as transparencies which could be enlarged optically to those scales. The data obtained from the Indian Remote Sensing Satellite IRS-1A, especially from its LISS II sensor with 36.5m resolution, is also immensely useful, as these are comparable to the spectral bands of Landsat TM bands 1 to 4. Therefore, LISS II FCC of bands 2+3+4 is also recommended for hazard mapping. The easy availability of the IRS-IA data products in the country at a much cheaper rate than the Landsat data product should be a boon to the potential researchers. The French SPOT satellite data products are much more costlier than the above two, but have far better resolution also. The multispectral mode (or MLA) of the imaging system records data in the following wavelengths : (1) 0.50-0.59, (2) 0.61-0.68, and (3) 0.79-0.89 micrometer, where the spatial resolution is 20 m. Hence the MLA FCC of bands 1+2+3 provides enormous information on the terrain. The panchromatic mode (PLA), on the other hand, works more like an aerial camera (wavelength 0.51-0.73 micrometer) and has a

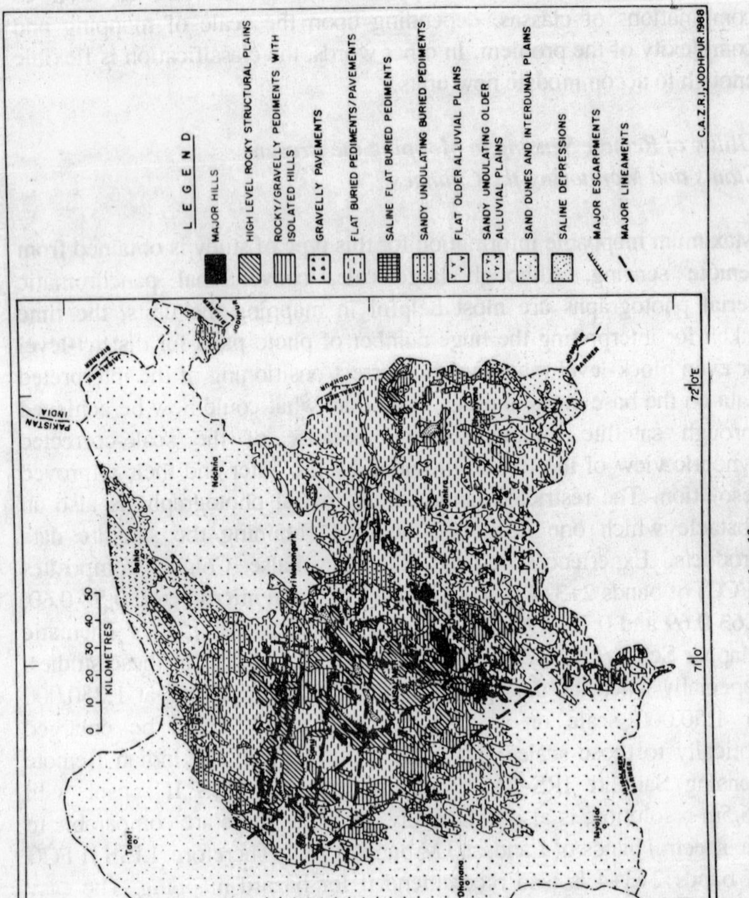

FIG. 8.3: Jaisalmer district: landforms.

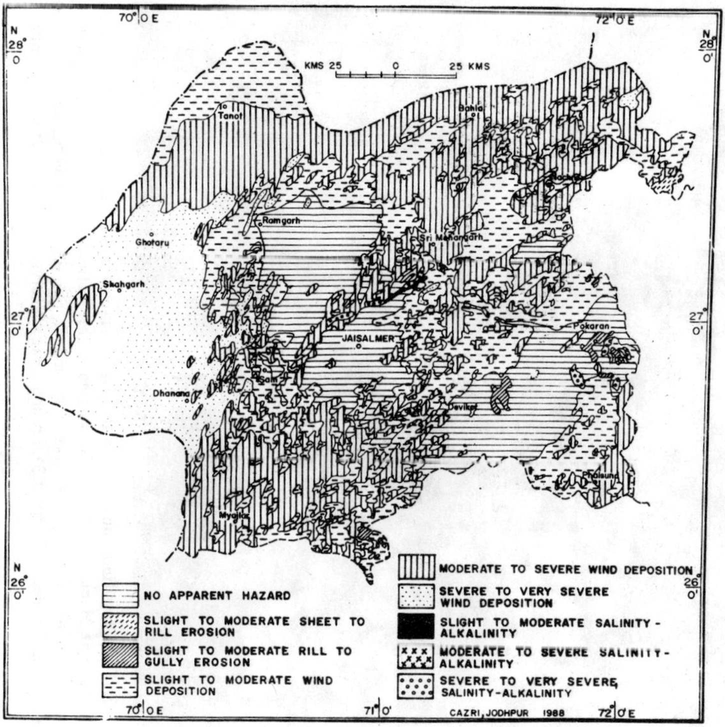

FIG. 8.4: Jaisalmer district: Erosional, depositional and salinity hazards.

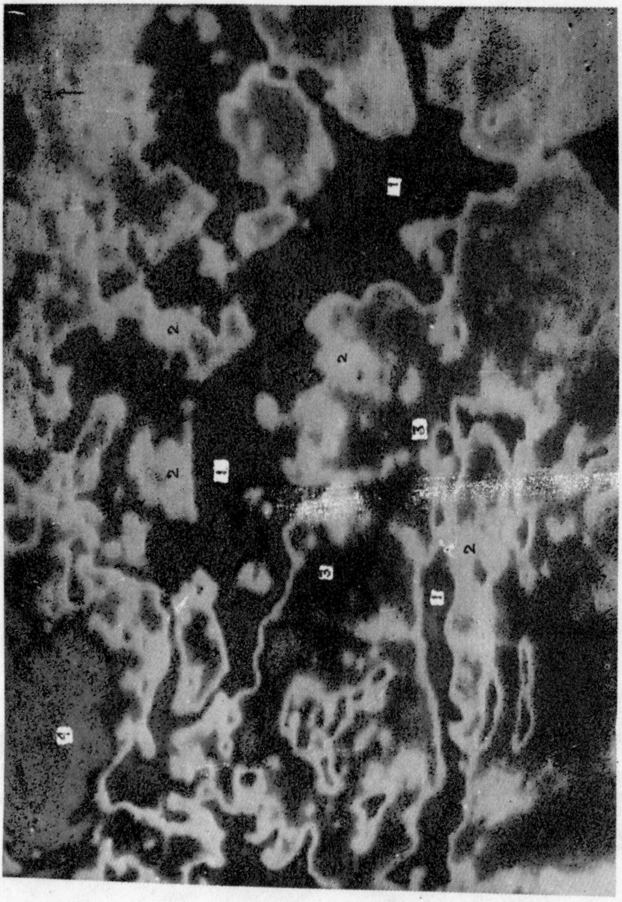

PLATE 8.3: Classified subscene of the Banni (raised mud flat) around Bherandiala, showing: (1) high, less saline Banni; (2) low, moderate to highly saline Banni; (3) Rann, without appreciable salt encrustation — usually wet; (4) Rann, with highly reflective salt encrustation — usually dry. Image generated from Landsat TM (band 3) through Image Analyser (original output in colour).

spatial resolution of 10 m. Repeated off nadir viewing of the ground by the SPOT sensors produces overlapping of images of the same area and hence, the image pairs obtained during a 26-days cycle are amenable to stereoscopic viewing.

The digital remote sensing techniques, using data products in the form of computer compatible tapes, have so far yielded much less reliable data in our desert than what an interpreter can decipher through visual interpretation of the photographic prints (Kar, et al, 1989). The digital remote sensing is also very much costly and has an adverse cost-benefit ratio for geomorphological mapping in large, heterogenous desert terrain. On the other hand, the use of an Image Analyser that performs density slicing of a single band image in black and white print, or part thereof, and assigns upto eight pseudo-colours to the narrow bands of grey tones chosen by the interpreter, yielded some useful results to the author (Plate 8.3).

The repeatitive coverage of the terrain by the above satellites do also help in monitoring the changes in spatial dimension of the hazards, both short-term (especially during the extreme events like floods, severe sandstorms, extreme droughts, etc.) and long-term. The assigning of any particular hazards class to an area is, however, subjective and depends to a great extent on the investigator's ability to perceive the hazards and their intensity in broader perspective of the desert environment and processes.

When the map on hazards is superimposed on a landform map and maps on other physical aspects of the land, the major problems within an area and their causes are clearly perceived (Figs. 8.3 and 8.4).

REFERENCES

Aggarwal, S. C., 1957 : *Pachbadra and Didwana salt sources,* Govt. of India Press, Delhi, p. 206.

Allchin, B., Goudie, A. and Hegde, K. T. M., 1978 : *The prehistory and palaeogeography of the Great Indian Desert,* Academic Press, London, p. 370.

Biswas, S. K., 1971 : Note on the Geology of Kutch, *Quart. J. Geol. Min. Met. Soc. India,* 43, pp. 233-35.

———, 1974 : Landscape of Kutch — A morphotectonic analysis, *Ind. J. Earth Sc.,* 1, pp. 177-90.

Burnes, A., 1835 : Memoir on the eastern branch of the river Indus, *Trans. Asiatic Soc.,* 3, pp. 550-88.

Chepil, W. S., Siddoway, F.H. and Ambrust, D.V., 1962 : Climatic factor for estimating wind erodibility of farm fields, *J. Soil and Water Cons.,* 17, pp. 162-65.

Dassarma, D. C., 1986 : Neotectonism in Rajasthan — its manifestations and effects, *Proc. Int. Symp. Neotectonics in South Asia,* Survey of India, Dehra Dun, pp. 282-88.

Dhir, R. P., 1989 : Wind erosion in relation to landuse and management in Indian arid zone, *Int. Symp. Managing Sandy Soils,* Jodhpur, Abstracts, pp. 572-75.

Dhir, R. P., Kolarkar, A. S., Sharma, K. D., Vangani, N. S., Saxena, S. K., Sen, A. K., Ramakrishna, Y. S., Murthy, K. N. K., Singh, N. and Tak, B. L., 1982 : *July 1979 flash flood in the Luni — a case study,* CAZRI Tech. Bull. 6, p. 36.

Fairbridge, R. W., 1968 : Quaternary period, In, Fairbridge, R. W. (ed), *Encyclopaedia of Geomorphology,* pp. 912-31.

Ghose, B., 1964 : Geomorphological aspects of the formation of salt basins in western Rajasthan, *Proc. Symp. Problems of Indian Arid Zone,* CAZRI, pp. 79-83.

Ghose, B., Kar, Amal and Husain, Z., 1979 : The lost courses of the Saraswati river in the Great Indian Desert — new evidence from Landsat imagery, *Geog. J.,* 145, pp. 446-51.

Ghose, B., Singh, S. and Kar, Amal, 1977 : Desertification around the Thar — a geomorphological interpretation, *Ann. Arid Zone,* 16, pp. 290-301.

———, 1982 : Some geomorphic aspects of salinity hazard in Rajasthan desert, India, *Proc. Workshop on the problems of the deserts in India,* Geol Surv. India Misc. Publ. 49, pp. 11-14.

Goudie, A., Allchin, B. and Hegde, K. T. M., 1973 : The former extensions of the Great Indian Sand Desert, *Geog. J.,* 139, pp. 243-57.

Kar, Amal, 1981 : Land, man and desert degradation — geomorphic problems of agriculture in western Rajasthan. In, Noor Mohammad (ed.), *Perspectives in agricultural geography,* Vol. 2, Concept, New Delhi (Ist edn.), pp. 77-87.

———, 1983 : Morphology and evolution of slopes in the upper Luni basin, Rajasthan, *Ann. Arid Zone,* 22, pp. 367-75.

———, 1986 : Physical environment, human influences and desertification in Pushkar-Budha-Pushkar lake region, India, *The Environmentalist,* 6, pp. 227-32.

———, 1987a : Origin and transformation of longitudinal sand dunes in the Indian desert, *Zeit. for Geomorph.* 31, pp. 311-37.

———, 1987b : Nature, man and issues in development — lessons from the upper Luni basin, In, Sharma, H.S. and Sharma, M.L. (eds.), *Environmental design and development,* Scientific Publishers, Jodhpur, pp. 185-92.

———, 1988a : Possible neotectonic activities in the Luni-Jawai plains, Rajasthan, *J. Geol. Soc. India,* 32, pp. 522-26.

———, 1988b : Remote sensing as an aid for identification and mapping of present-day processes and natural hazards in the Indian desert, In, Sahai, B. (ed.), *Remote Sensing in Agriculture,* SAC, Ahmedabad, pp. 216-21.

———, 1989a : Evidence for neotectonism from the Indian desert, In, Singh, S. and Tiwari, R. C. (eds.), *Geomorphology and environment,* Allahabad Geog. Soc., Allahabad, pp. 300-10.

———, 1989b : Evaluation of present-day geomorphic processes in the Indian desert for assessing land degradation and natural hazards, *Ibid.,* pp. 643-54.

———, 1989c : Sand dunes and their mobility in Jaisalmer district, *Second Conf. Indian Institute of Geomorphologists,* Jaipur [to be published in Dixit, K. R. and Kaul, M. N. (eds.), *Geomorphological Diversity*].

———, 1989d : Terrain characteristics of Jaisalmer district, *Geog. Rev. India*, 51, pp, 48-59.

———, 1989e : Drainage desiccation, water erosion and desertification in north India, *Workshop on Tackling Soil Erosion, Desertification and Drought-related Problems for Commonwealth African Countries,* CAZRI, Jodhpur.

———, 1990a : A stream trap hypothesis for the evolution of some saline lakes in the Indian desert, *Zeit. Fur. Geomorph.* 34(1), pp. 37-47.

———, 1990b : Megabarchanoids of the Thar — their environment morphology and relationship with longitudinal dunefield, *Geog. J.*, 156(1).

———, Morphology and vulnerability of the Indian desert landforms, *Geog. Rev. India.* (forthcoming).

———, Geomorphology of the Thar desert in Rajasthan, In, Sharma, H. S. and Sharma, M. L. (eds.), *Changing Scenario of Geography of Rajasthan,* Kuldeep Prakashan, Ajmer (in press).

Kar, Amal and Ghose, B., 1984 : The Drishadvati river system of India — an assessment and new findings, *Geog. J.*, 150, pp. 221-29.

Kar, Amal and Joshi, D. C., 1989 : Sand movement and control of aeolian hazard, *Expert papers, Regional meeting of experts, local level Govt. officials and NGO's on desertification control in the Asia and Pacific Region*(ESCAP/UNDP/ICAR), AKDJ 1 — AKDJ 19.

Kar, Amal and Shankarnarayan, K. A., 1984 : Some environmental considerations for the management of surface water in the upper Luni basin of Rajasthan, India, *Ind. J. Geog.* 14, pp. 33-42.

Kar, Amal and Singh, S., 1989 : Genesis, distribution and classification of aeolian hazards in the Indian desert, *Int. Symp. Managing Sandy Soils*, Jodhpur, Abstract, pp. 546-50.

Kar, Amal, Singh, S., Vangani, N. S., Dubey, S. C. and Ghose, T. K., 1989 : Application potentials of IRS-1A data for mapping hydrogeomorphic features in the Indian desert, *Regional Workshop on IRS Mission and its application potentials,* RRSSC, Jodhpur, pp. 54-58.

Krishnan, M. S., 1952 : *Geology of India and Burma,* Higginbothans, Madras.

Lal, B. B., 1979 : Kalibangan and Indus Civilization, In, Agrawal, D. P. and Chakrabarti, D. K. (eds.), *Essays in Indian protohistory,* B. R. Publ., Delhi, pp. 65-97.

Mann, H. S. and Chatterji, P. C., 1978 : Effect of landuse change on the groundwater conditions in canal irrigated area of Western Rajasthan and their management, *Proc. Symp. Land and Water Management in the Indus Basin* (India), 1, pp. 252-68.

Shankarnarayan, K. A. and Kar, Amal. (eds.), 1983 : *Upper Luni Basin — an integrated analysis of natural and human resources for development planning,* CAZRI, p. 577.

Singh, G., Joshi, R. D., Chopra, S. K. and Singh, A. B., 1974 : Late Quaternary history of vegetation and climate of the Rajasthan desert, India, *Phil. Trans. Royal. Soc.*, London, 267 B, pp. 467-501.

Singh, S., Ghose, B. and Kar Amal, 1978 : Geomorphic changes as evidence of palaeoclimate and desertification in Rajasthan desert, India (Luni development block — a case study), *Man & Environ.*, 2, pp. 1-13.

Singh, S. and Kar Amal, 1988 : Geomorphological appraisal of arid waterlands by using remote sensing techniques. In, Shankarnarayan, K. A. (ed.), *Wastelands development and their utilisation,* Scientific Publishers, Jodhpur, pp. 62-86.

Skidmore, E. L., 1988 : Wind erosion, In, *Soil erosion research methods,* Soil and water Cons. Soc., USA, pp. 203-33.

Wadia, D. N., 1960 : The post-glacial desiccation of Central Asia — evolution of the arid zone of Asia, *Ind. Nat. Sc. Acad.*, Manograph. 1.

Woodruff, N. P. and Siddoway, F. H., 1965 : A wind erosion equation, *Proc. Soil Sc. Soc. Amer.*, 29, pp. 602-08.

Yaalon, D. H. and Ganor, E., 1966 : The climatic factor of wind erodibility and dust blowing in Israel, *Israel J. Earth., Sc.* 15, pp. 27-32.

9

WEATHER-BASED FORECASTS AND ITS IMPLICATIONS

W. J. MAUNDER

GIVEN SUFFICIENT lead time, the value of accurate weather-based forecasts of New Zealand's pastoral production are wide-ranging, involving many activities beyond the farm gate, but the key factor in optimising the value of these forecasts is the need for more realisation within the industry that weather variations can and indeed must be taken into account. In the United States, this is now taken place, indeed, in 1977 the U.S. Secretary of Agriculture issued a ruling that ". . .never again do I want to see an economic report come out of the (U.S.) Department of Agriculture that assumes average weather".

This average weather misconception is regrettably still a feature of many aspects of the pastoral production industry of New Zealand. Indeed, although many informed decision-makers within the industry are very much aware of the importance of weather variations, others seem to be quite content to go from one season to the next with an almost complete disregard to the effects of weather variations on productivity, transportation and marketing, this disregard affecting the farming community and the nation.

The Economic Service of the New Zealand Meat and Wool Boards is, however, one group of decision-makers within the industry which is very aware of the importance of weather variations. Thompson and Taylor (1975) of that Service state, for example:

> Climatic variations affect both the seasonal volume of production and its movement to markets, with implications for transport and sales which must be co-ordinated world-wide. Unpredictability of supply at whatever level disrupts the smooth flow of product from

farm gate to consumer, increasing costs to all involved. Weather effects on flow patterns of major products within seasons can be significant, with implications for the scheduling of auctions in the case of wool and for meat, and the arrangement of shipments to export markets. Apart from the marketing aspects, steady product flows enable better use of transport and storage facilities and contribute to reducing costs.

Production, Processing and Transportation

The fluctuations in New Zealand's pastoral production from season to season are well illustrated in Tables 9.1 and 9.2, and as Troughton (1977) comments:

> Farm production is inherently seasonal because of climatic and biological constraints on plant and animal growth, and this introduces a distinctive and specialised component into the management of our agricultural system. A major problem is to reconcile the highly seasonally-dependent production with processing and market demand. This approach to management introduces the need to consider the whole agricultural chain from the farm to the marketplace as one integral system.

TABLE 9.1: **Fluctuations* in New Zealand meat and wool production, 1964-65 — 1976-77**

Season	Beef (1)	Mutton (1)	Lamb (1)	Wool (2)
1964-65	− 4.6	− 3.9	− 1.8	+ 1.4
1965-66	+ 7.5	−11.6	+ 4.4	+11.3
1966-67	+ 2.2	+23.7	+ 4.7	+ 2.2
1967-68	+15.9	+14.6	+ 5.4	+ 2.4
1968-69	+ 9.4	− 7.5	+ 3.5	+ 0.6
1969-70	+ 4.8	0.0	− 0.1	− 1.3
1970-71	+ 1.4	+ 2.5	− 1.1	+ 1.8
1971-72	+ 4.7	− 4.5	+ 5.5	− 3.6
1972-73	+ 8.7	+ 9.8	−10.1	− 4.1
1973-74	−10.9	−10.4	−10.7	− 7.8
1974-75	+27.9	−24.4	+11.1	+ 3.3
1975-76	+34.5	+ 1.5	+10.9	+ 6.0
1976-77	−18.2	+ 6.1	− 4.4	− 3.0

* percentage change from previous season.
(1) Export season October 1 to September 30.
(2) Export season July 1 to June 30.

Source: Compiled from information supplied by the New Zealand Meat Board and the New Zealand Wool Board.

TABLE 9.2: Milkfat processed by dairy factories in New Zealand

(million kg)

Season	Total Production	June-Dec.	Jan.-May
1968-69	267	156	111
1969-70	243	162	81
1970-71	243	151	93
1971-72	258	154	104
1972-73	247	161	85
1973-74	228	152	76
1974-75	244	149	95
1975-76	268	159	109
1976-77	275	170	105
1977-78	251	169	82

Source: Compiled from information supplied by the Farm Production Branch of the New Zealand Dairy Board.

The month-to-month variations in the meat, wool and dairy industries are also extremely important, and this is shown in Tables 9.3, 9.4, 9.5 and 9.6. In the case of meat production, Taylor (1977a) notes that one of the main requirements for an efficient production, processing and marketing chain is the accurate forward estimation of the volume of produce which the chain is required to handle. The pastoral industry, he says, is characterised by wide variations in the total volume which comes forward for processing each year, and he comments:

> Thus it becomes essential to establish a methodology for accurately estimating the volume of meat available for processing in advance, and adjusting this within the season as conditions change. . . . As a country so dependent on exports can we afford to see our major export-earner function at anything less than peak efficiency through the whole pasture to market chain?

During the season 1977-78 the total value of pastoral-based exports from New Zealand was over $ 2,400 million, representing about 70 per cent of the total value of exports. A considerable proportion of the product is transported by container ships, the costs of operation of which are very high (typical costs are $ 15,000 a day while in port, and $ 25,000 a day while at sea). Freight costs are based among other things on : (a) the number of days of voyage, (b) the costs per day, and

(c) the load factor; therefore, any system which can increase the overall efficiency of the whole transportation industry of New Zealand pastoral production is worthwhile, not only to the farmer but also to the nation.

TABLE 9.3: Monthly slaughterings of lamb for export

(000)

Month	Season 1975-76	Season 1976-77	Diff.*	Season 1977-78	Diff.**
Nov.	2160	1617	− 543	1402	− 215
Dec.	3554	4675	+1121	3398	−1277
Jan.	4710	4264	− 466	4739	+ 475
Feb.	3602	3444	− 158	4527	+1083
Mar.	3666	4276	− 610	3773	− 503
Apr.	3313	2681	− 632	2992	+ 311
May	2561	2621	+ 60	2559	− 62

* 1976-77 — 19754-76.
** 1977-78 — 1976-77.

Source: Compiled from information supplied by the New Zealand Meat and Wool Boards' Economic Service.

TABLE 9.4: Cumulative slaughterings of lambs for export

(000)

To end of month	Season 1975-76	Season 1976-77	Diff.*	Season 1977-78	Diff.**
Dec.	5868	6435	+567	4868	−1567
Jan.	10579	10699	−120	9608	−1091
Feb.	14181	14143	− 38	14135	− 8
Mar.	17847	18419	+572	17908	− 511
Apr.	21166	21100	− 66	20899	− 101
May	23722	23721	− 1	23458	− 263
Average weight kg/head	13.7	13.3	−0.4	12.8	−0.5
Meat production † (million tonnes)	325	315	−10	300	−15

* 1976-77 — 1975-76.
** 1977-78 — 1976-77.
† October 1 to May 31.

Source: Compiled from information supplied by the New Zealand Meat and Wool Boards' Economic Service.

TABLE 9.5: Monthly disposal of wool

(million kg)

Month	Season			Season	
	1975-76	1976-77	Diff.*	1977-78	Diff.**
Oct.	22.9	16.2	−6.7	15.6	− 0.6
Nov.	24.1	23.8	−0.3	30.1	+ 6.3
Dec.	41.2	42.8	+0.6	40.1	− 2.7
Jan.	38.1	42.2	+4.1	40.1	− 2.1
Feb.	35.5	40.6	+5.1	43.9	+ 3.3
Mar.	42.7	43.3	+0.6	31.0	−12.3
Apr.	35.8	27.8	−8.0	39.9	+12.1
May	27.2	24.5	−2.7	24.4	− 0.1
June	18.4	18.1	−0.3	19.1	+ 1.0
Total†	313.7	310.0	−3.7	313.8	+ 3.8

* 1976-77 — 1975-76.
** 1977-78 — 1976-77.
† October to June.

Source: Compiled from information supplied by the New Zealand Wool Board.

TABLE 9.6: Milkfat processed by dairy factories in New Zealand in March, and percentage changes from March — March

Year	Production million kg	Value* $ million	Production % of previous year
1968	19.5	48.7	79
1969	25.6	64.0	131
1970	13.8	34.5	54
1971	20.9	52.3	151
1972	21.7	54.2	104
1973	16.7	41.8	74
1974	16.8	42.0	100
1975	20.6	51.5	124
1976	24.1	60.3	114
1977	22.8	57.0	96
1978	17.4	43.5	76

* Based on a value of $2.50/kg. which is the value in May, 1978, if milkfat is converted into cheese and exported.

The weather factor in this efficiency aspect of the problem, particularly in the effect of weather on seasonal production, is very significant. However, the real effects of weather variations tend to be hidden from the official concern expressed by producer boards. For example, the New Zealand Meat Producers' Board in their 55th Annual Report (1977) stated:

> "With the rapidly-increasing container traffic, there is a need for more communication between the various parties to ensure maximum and efficient use of facilities and to avoid unnecessary delays. . . . The high daily running costs of container vessels make it essential that the quickest possible turnround be achieved if we are to avoid paying additional and unnecessary costs. This year our freight bill for meat alone is around $ 180 million. Rates charged to us are subject to many influences. . .and we are continuing to pursue ways of holding costs within New Zealand to minimise the effect of freight increases."

Costs Beyond the Farm Gate

New Zealand's success as an exporter of meat and wool has been highly dependent upon her ability to produce high quality products at a relatively low cost. This ability was maintained over a long period of time but the situation is now very different, in that the costs "beyond the farm gate" have reached critical levels.

Ward (1978) notes that . . . "these costs now severely inhibit New Zealand's ability to succeed as an exporting nation and have almost more impact than the level of overseas prices". An indication of these costs is given in Tables 9.7, 9.8, 9.9 and in some pertinent comments, Taylor (1977b) states:

> The costs incurred on transporting, processing and marketing of a 14.2 kg lamb in 1975 was $ 9.25 per head which has gone to $ 13.51 in 1977 showing an increase of 46 per cent over a period of just two years (Table 9.8). In percentage terms the largest increase over this period occurred in the areas associated with transport, i.e. farm to works, and f.o.b. ex-hooks Smithfield.

It is considered that the meteorological input into minimising these costs is a significant factor. Indeed, if the close liaison between the meteorological system and the pastoral producing industry is strengthened, the significance of weather variations to the pastoral industry would become fully realised. Some aspects of the weather-based pastoral production models currently in use are now discussed.

TABLE 9.7: Major components of United Kingdom lamb export price/cost structure*

Component	Season		
	1974-75	1975-76	1976-77
Farm gate	38	55	68
Transport to works	2	2	2
Schedule (season average)	41	57	70
Works to f.o.b. (mostly processing charges)	21	25	29
f.o.b.	66	85	101
Freight, insurance UK levies, landing cost	44	55	67
Ex-hooks, Smithfield	111	140	169
Total costs, farm gate to Smithfield	73	85	101
Additional costs as %	192	155	148

* Cents per kg of P.M. grade.
Source: Reserve Bank of N.Z. (per M.J. Walsh).

TABLE 9.8: Estimated charges on N.Z. lamb from farm-gate to the U.K.

(*$ per lamb*)

Item	January		
	1965	1975	1977
Transport to works	0.13	0.25	0.33
Works to f.o.b.	1.00	3.29	3.85
f.o.b. to Smithfield	1.59	5.71	9.33
Total costs	2.72	9.25	13.51
Bare meat value at Smithfield	6.77	15.75	25.92
Costs as % of meat value	41	59	52

Source: N.Z. Meat and Wool Boards' Economic Service (per N.W. Taylor).

Weighted Weather Indices and National Production

Appropriate production data such as the distribution of sheep, dairy cows and beef cattle in various areas have been used in association with weather information such as rainfall, temperature and days of deficit from about 100 stations to develop weighted weather indices. The methodology used in such analyses is given in Maunder (1972), and recent appraisals of the use of weighted weather information in national planning are given in Maunder (1977 a, c, d).

TABLE 9.9: Major components wool of export price/cost structure*

Component	Season		
	1974-75	1975-76	1976-77
Farm gate	84	146	206
Transport, insurance, brokers' charge, Wool Board levy	8	11	14
Auction price (season average)	92	157	220
Delivery and pre-shipment charges, buyers commission	4	6	9
Estimated f.o.b.	96	163	229
Ocean freight and marine insurance	11	18	19
Estimated c.i.f.	106	181	248
Port charges and transport to mill	2	3	3
Estimated landed cost overseas	108	184	252
Total costs, farm gate to overseas mill	24	38	46
% Additional costs as %	22	21	18

* Cents per kg of greasy wool.

Source: Reserve Bank of N.Z. (per M.J. Walsh).

As far as the pastoral industry is concerned, the weighted number of days of deficit is a convenient and useful way of condensing a vast amount of weather information into a compact index. The number of days of deficit is calculated from the balance between the daily rainfall data and estimates of the daily potential evapotranspiration, assuming a potential rate of evapotranspiration until 75 mm. of water has been withdrawn from the soil.

The weighted number of days of deficit for New Zealand for the 29 seasons 1949-50 to 1977-78 using three different weightings are shown in Table 9.10. These data are for New Zealand as a whole, and can, therefore, be said to represent the national sheep farm, the national dairy farm and the national beef cattle farm respectively. The indices show just how variable the New Zealand pastoral climate is; noteworthy are the very high values in the recent drought seasons of 1972-73, 1973-74 and 1977-78, which are in marked contrast to the green era of the mid-1960s. For example, the New Zealand indices weighted by the distribution of dairy cows give a total index of 15.2 days for the three consecutive seasons, 1964-65, 1965-66, 1966-67, this being in very marked contrast to the average index during the six seasons, 1972-73 to 1977-78, of 30.7 days, and the total index during the two consecutive drought seasons of 1972-73 and 1973-74 of 80.8 days.

TABLE 9.10: Weighted* number of days of water deficit — New Zealand

Season	June-May			March-May		
	Sheep	Cows	Cattle	Sheep	Cows	Cattle
1949-50	41.3	40.3	43.3	19.3	17.9	19.4
1950-51	19.7	13.0	15.2	7.2	5.6	5.1
1951-52	34.6	29.4	31.2	22.7	23.1	20.1
1952-53	16.2	12.7	14.0	7.2	10.4	9.5
1953-54	39.2	36.9	14.1	6.3	2.6	4.4
1954-55	38.2	35.0	35.9	12.5	12.3	11.4
1955-56	42.9	20.8	33.2	13.5	9.2	10.6
1956-57	26.7	23.0	28.7	6.6	5.7	8.8
1957-58	20.7	19.5	29.9	5.5	2.8	8.3
1958-59	22.7	4.7	13.9	4.0	0.5	2.1
1959-60	34.2	13.7	27.0	10.1	3.4	7.1
1960-61	19.5	19.0	17.0	5.7	9.7	6.9
1961-62	35.1	22.4	31.9	3.5	0.7	1.9
1962-63	29.8	23.4	28.3	11.8	10.5	12.0
1963-64	42.4	26.4	39.2	11.7	5.2	10.8
1964-65	18.5	4.8	12.6	3.8	1.0	2.0
1965-66	19.3	4.6	14.1	8.0	1.5	4.8
1966-67	26.6	5.8	14.9	9.9	3.8	8.1
1967-68	32.3	29.1	35.2	13.4	13.4	16.8
1968-69	32.7	12.8	24.1	16.9	9.4	14.2
1969-70	33.9	34.2	30.9	9.9	7.8	7.6
1970-71	38.8	16.4	26.7	13.3	3.6	9.2
1971-72	31.0	19.8	24.5	6.4	2.3	3.4
1972-73	59.7	45.8	52.8	17.9	13.5	15.0
1973-74	40.1	35.0	37.2	8.7	4.9	7.7
1974-75	30.3	17.1	25.0	6.2	4.6	6.6
1975-76	36.9	20.8	25.3	22.2	16.3	17.1
1976-77	27.2	24.1	22.8	17.9	15.6	14.9
1977-78	55.2	41.3	51.6	23.0	24.9	24.1

* Weighted by the distribution of sheep, dairy cows and beef cattle as shown.

The use of this weighted weather information in weather-based production models is now discussed, firstly for national dairy production, and secondly for national wool production. Real-time production forecasts based on these models have been made on a routine basis since 1972-73 for dairy production, and since 1977-78 for wool production.

Dairy Production Forecasts

The current weather-based dairy production model (for details see

Maunder, 1974, 1977b) is based on the differences in the weighted number of days of water deficit, the weighted rainfalls, and the weighted mean temperatures. The resulting differences are then used in the model to predict dairy production as a percentage change from the corresponding month(s) in the previous season. Typical days of deficit data used in these models are given in Tables 9.11 and 9.12, and typical dairy production data are given in Table 9.13.

TABLE 9.11: Weighted* number of days of water deficit — New Zealand

Season	Jan.	Feb.	Mar.	Apr.	Jan.-Apr.
1964-65	2	1	0	0	3
1965-66	0	1	1	0	2
1966-67	1	1	2	2	6
1967-68	2	14	11	1	28
1968-69	2	2	7	4	15
1969-70	6	18	7	1	32
1970-71	4	6	1	3	14
1971-72	5	12	2	0	19
1972-73	8	20	10	4	42
1973-74	15	13	4	1	33
1974-75	4	7	4	1	16
1975-76	1	2	15	1	19
1976-77	1	8	13	2	24
1977-78	7	15	18	7	47

* Weighted by the distribution of dairy cows.

TABLE 9.12: Weighted* water deficit indices** — New Zealand

Season	Jan.	Feb.	Mar.
1970-69	+5	+16	+2 *
1971-70	−2	−13	−6
1972-71	+1	+6	+0
1973-72	+3	+9	+8
1974-73	+7	−7	−6
1975-74	−11	−6	+0
1976-75	−3	−4	+11
1977-76	−0	+8	−0
1978-77	+6	+7	+5

* Weighted by the distribution of dairy cows.

** Differences between season in days (i.e. January 1978 — January 1977 + 6 days).

TABLE 9.13: Predicted New Zealand dairy production in April, 1978 as a percentage of April, 1977, using January weather information

Predicting equation:

$y = 106.3 - 0.72\,x_1 - 0.08\,x_2 + 0.33\,x_3$

where y = Milkfat production expressed as a % of the same month last season (i.e. April, 1978 as % of April, 1977).

x_1 = difference in the deviation of the weighted* mean temperatures (tenths of degree) from the 1941-70 normal, between one *January* and the preceding *January* (i.e. January, 1978 — January, 1977).

x_2 = as for x_1, but differences in the weighted* number of days of soil moisture deficit (tenths of days).

x_3 = as for x_1, but differences in the weighted* rainfalls expressed as a % of the 1941-70 normal.

Prediction for April, 1978, as % of April, 1977:

y = 106.3 – 0.72 (+14*) – 0.08 (+62**) + 0.33 (–61***)

* January, 1978 = +0.8°C; January, 1977 = –8.6°C
Therefore difference = +1.4° = +14 units

** January, 1978 = 7.0 days; January, 1977 = 0.8 days
Therefore difference = +6.2 days = +62 units

*** January, 1978 = 41%; January, 1977 = 102%
Therefore difference = –61 units

i.e. y = 106.3 – 10.1 – 5.0 – 20.1
= 71.1

Therefore: Predicted New Zealand dairy production in April, 1978, as a % of the April, 1977, production is 71%.

Note: This prediction model used weather information only up to January, 1978, for predicting production in April, 1978. The actual production in April, 1978 (which was not known until late May) was 7.2 m. kg. of 46% of the April, 1977, production.

* Weighted by the distribution of dairy cows.

In order to quantity these relationships various regression models are used:

$$Y_n = a + bx_{(n-3)} + cx_{(n-2)} + dx_{(n-1)} + ex_{(n)} \quad (1)$$
$$Y_n = a + bx_{(n-3)} + cx_{(n-2)} + dx_{(n-1)} \quad (2)$$
$$Y_n = a + bx_{(n-3)} + cx_{(n-2)} \quad (3)$$
$$Y_n = a + bx_{(n-3)} \quad (4)$$

In practice it has been found that equation (2) is the most useful, in that it allows a prediction of production differences in month (n), say March, using only the weather difference data for months (n^{-1}), (n^{-2}), and (n^{-3}), i.e. February, January and December. Further, if the February weather conditions can be forecast the actual prediction for, say, March could be done as early as mid-January. Similarly, equations (3) and (4) can be used, giving additional lead time. Indeed, for many purposes, a reasonably successful forecast using equation (4), i.e. forecasting April production, using only January weather information, may well be of greater value than a much more accurate forecast given at the beginning of the month of production.

The prediction equation and resultant predictions for April, 1978, are given in Table 9.13, and this shows that using equation (4) (that is, using only January weather differences to predict April production differences), the predicted New Zealand dairy production in April, 1978, was 71 per cent of the April, 1977, production, whereas the actual April, 1978, production was 46 per cent. However, since the actual production of 46 per cent was not known until late May, 1978, the predicted production of 71 per cent, although a little optimistic, did give about 15 weeks' lead time to the decision-makers of the impending dramatic fall in production.

The importance of providing production forecasts with sufficient lead-time to enable appropriate decisions (transportation, marketing, etc.) to be taken must be emphasised, and a summary of the progressive forecasts made during the 1977-78 season is given in Table 9.14. The dramatic drop in production in February to April, 1977-78, compared with February to April, 1976-77, as shown in Table 9.14, was well anticipated (e.g. with an 11 weeks' lead-time the March drop in production was predicted to be 69 per cent of March, 1977, compared with an actual drop of 76%).

The significance of the 8 to 15-weeks lead-time predictions is well illustrated if one considers the cumulative production (actual and forecast) over the season. Table 9.15 gives such information for the 1977-78 season. It is considered that such weather-based forecasts are of direct benefit to decision-makers within the dairy industry, and if correctly applied, could lead to a considerably more efficient system beyond the farm gate.

In this regard it is pertinent to note the remarks of a Dairy Board representative (M.G. Boyer) in 1975 made while speaking at a Symposium on Meteorology and Food Production:

> How do we put a cost-benefit figure on the advantages that arise from better longer-range dairy production forecasts. The benefits include not overcommitting ourselves in sales, and not ordering shipping, containers and packaging materials not needed in seasons

when production tails off rapidly. . . . In years when storage is fairly full an accurate forecast of a dry autumn, would mean relief from storage problems. . . . Conversely, in a good season a slightly longer-term forecast will help us to organise things so as not to undersell the season's produce, or under-order transport, packaging and storage.

TABLE 9.14: Predicted milkfat production in New Zealand during 1977-78 season, with lead-time before actual production is known

(in weeks)

Month of prediction †	Month Predicted													
	Oct.		Nov.		Dec.		Jan.		Feb.		Mar.		Apr.	
	%	LT	%	LT	%	LT	%	LT	%	LT	%	LT	%	LT
Jul.	101*	(15)*												
Aug.	103	(11)	103	(15)										
Sep.	101	(8)	105	(11)	105	(15)								
Oct.	106	(3)	104	(8)	105	(11)	102	(15)						
Nov.			102	(3)	103	(8)	99	(11)	104	(15)				
Dec.					104	(3)	102	(8)	107	(15)	108	(15)		
Jan.							101	(3)	76	(8)	69	(11)	71	(15)
Feb.									74	(3)	63	(8)	68	(11)
Mar.											82	(3)	57	(8)
Apr.													55	(3)
Actual production	98		102		99		95		81		76		46	

† End of months shown.

* Prediction was made at the end of July, 1977, for the New Zealand October, 1977, milkfat production to be 101 per cent of the actual production in October, 1976. The actual production was not known until late November, giving a lead-time of 15 weeks.

Wool Production Forecasts

Wool production forecasts have recently been made using a weather-based model similar to that already discussed for forecasting dairy production. The models used are continually being modified and improved, but a summary of the model used in June, 1978, to forecast the wool production per hectare for 1978-79 (as a percentage of the 1977-78 season) is given in Table 9.16. This indicates that using weather information up to May, 1978, the predicted wool production per hectare for the 1978-79 season (that is, the season following the weather used in the model) is 97.8 per cent of the 1977-78 production, the main cause of this 2.2 per cent decrease being the 1978 autumn drought.

TABLE 9.15: Cumulative predicted and actual milkfat production in New Zealand during 1977-78

(in million kg)

Month of prediction †	Known production 77-78	Predicted production 77-78	Known † predicted production A	Actual production 77-78 B	Actual production 76-77 C	A/C (%)	B/C (%)
Oct.	Oct. —	Oct.-Jan. 159.6	159.6	Oct.-Jan. 151.0	153.0	104	99
Nov.	Oct. 39.5	Nov.-Feb. 142.0	181.5	Oct.-Feb. 172.6	179.7	101	96
Dec.	Oct.-Nov. 80.1	Dec.-Mar. 128.3	208.4	Oct.-Mar. 190.0	202.5	103	94
Jan.*	Oct.-Dec. 117.7	Jan.-Apr. 82.2	199.9	Oct.-Apr. 197.2	218.1	92	90
Feb.	Oct.-Jan. 151.0	Feb.-Apr. 44.8	195.8	Oct.-Apr. 197.2	218.1	90	90
Mar.	Oct.-Feb. 172.6	Mar.-Apr. 27.5	200.3	Oct.-Apr. 197.2	218.1	92	90
Apr.	Oct.-Mar. 190.0	Apr. 8.6	198.6	Oct.-Apr. 197.2	218.1	91	90

† End of month shown.

* At the end of January, the known production (October to December) was 117.7 million kg. the predicted production (January to April) was 82.2 million kg. giving a "known plus predicted" production for the October-to-April period of 199.9 million kg. This is 92 per cent of the production for the same period of the previous season, and compares with 90 per cent which was the actual percentage — not known until late May, 1978.

Predictions using the same model for previous seasons are shown in Table 9.17, and this indicates that reasonably accurate forecasts of wool production can be given before the start of a season which gives up to 14 months' lead-time before the actual production is known. While the variations (both predicted and actual) may appear to be small, it must be emphasised that even a two per cent variation from season to season, is equivalent to a variation in production of six million kilogram of wool, worth at least $12 million and requiring a carrying capacity of about one container ship.

TABLE 9.16: Predicted New Zealand wool production in 1978-79 as a percentage of 1977-78, using weather information up to May, 1978

Predicting equation:

$$y = 101.4 + 0.038x_1 - 0.022x_2 + 0.022x_3 - 0.048x_4$$

Where y = wool production per hectare expressed as a % of the previous season (*i.e.* 1978-79 as % of 1977-78).

x_1 = difference in the weighted + number of days (tenths of days) of soil moisture deficit for the period June-to-October between seasons. (*i.e.* June-to-October, 1977, compared with June-to-October, 1976).

x_2 = difference in the deviation of the weighted* cumulative mean temperatures (tenths of degree) from normal for the period June-to-October between seasons.

x_3 = as for x_1 but period November-January.

x_4 = as for x_1 but period February-May.

Prediction for 1978-79 as % of 1977-78:

$$y = 101.4 + 0.038(+5^{*}) - 0.039(+1^{**}) + 0.022(+134^{***}) - 0.048(+141^{****})$$

* June-to-October, 1977 = 0.5 days
June-to-October, 1976 = 0.0 days
Therefore difference = + 0.5 days = +5 units

** June-to-October, 1977 = −1.2°C (cumulative)
June-to-October, 1976 = −1.3°C (cumulative)
Therefore difference = +0.1°C = +1 unit

*** November, 1977 to January, 1978 = 16.3 days
November, 1976 to January, 1977 = 2.9 days
Therefore difference = 13.4 days = 134 units

**** February, 1978 to May, 1978 = 38.4 days
February, 1977 to May, 1977 = 24.3 days
Therefore difference = + 14.1 days = 141 units

i.e. $y = 101.4 + 0.2 - 0.0 + 3.0 - 6.8 = 97.8$

Therefore : Predicted New Zealand wool production per hectare in 1978-79 as a % of the 1977-78 production is 97.8.

Note : This prediction model used weather information only up to May, 1978, for predicting the total production in the 1978-79 season.

\+ Weighted by the distribution of sheep.

TABLE 9.17: New Zealand wool production: predicted and actual*

Date of prediction	Season	Predicted wool/ha	Actual Sheep population	/ha	Actual production /Sheep	Total
May 1973	1973-74	97	93	95	99	92
May 1974	1974-75	106	99	116	105	103
May 1975	1975-76	102	99	101	107	106
May 1976	1976-77	97	102	101	95	97
May 1977	1977-78	102	105		98	103
May 1978	1978-79	98				

* All data expressed as a percentage of the previous season. Actual data complied from information supplied by the Economic Service of the New Zealand Meat and Wool Boards, and the Department of Statistics.

Note: (1) Actual data on wool production per hectare is usually not available until at least 15 months after the end of the season. Thus the data for 1976-77 would be available in November, 1978.

(2) Data on the sheep population is for the beginning of the season (i.e. 30 June).

(3) Data on the actual wool production per sheep and the total actual wool production is known at the end of each season, usually with a delay of about one month. Thus as of May, 1978, when a weather-based forecast of the wool production per hectare was made for the 1978-79 season, the actual production was known for the 1976-77 season, and provisionally available for 1977-78, the final production for the 1977-78 season not being available until August, 1978.

Conclusion

The main factors influencing farmers' livestock investment response, according to Walsh (1978), can be summarised in the following approximate order of importance: (a) terms of exchange, (b) climatic conditions and biological factors, (c) technological change and innovation, (d) availability of resources of land, labour and credit, and (e) other factors influencing farmers' attitude and confidence.

Walsh further comments that the New Zealand Meteorological Service has developed a number of national climatic variables and that in recent years this information ". . .has been increasingly used for its high predictive power in forecasting livestock performance (milk production, wool weights, etc.) but not as yet, it appears, for assessment of livestock investment".

It is pleasing to note these comments of Walsh (who was writing in a publication of the Reserve Bank of New Zealand), but perhaps they should be extended to suggest that the overall weather conditions in

New Zealand during the past decade have *not* been conducive to creating confidence in the farming community. For example, reference to Table 9.7 shows very high numbers of days of deficit during three recent seasons (1972-73, 1973-74, 1977-78). These droughts, coupled with the ever increasing costs of farming, must be taken into account, for, in simple terms, the farming community appears to be less able to withstand the *increasing* impact that weather variations have on productivity and activities beyond the farm gate.

Taylor (1978) has commented indirectly on some aspects of this problem, and he states:

> . . .today nearly 43 per cent of the total Government research effort goes into the agriculture sector, and the bulk of the agricultural effort (83%) goes towards on-farm or production research. When the activities outside the farm gate. . .are absorbing such a high proportion of the final market value. . .a case can be made. . .for the allocation of a greater proportion of this research effort into the off-farm area.

The potential gains in the off-farm areas are indeed becoming more and more significant, and it is believed that some of these gains could be realised through a better understanding and application of weather information, which has been specifically tailored to meet national planning and production activities.

The guarantee of a regular supply of new Zealand's pastoral products to an increasingly food-needy and food-conscious world is important. This regular supply is dependent to a very large extent on the actual pastoral production, and it is very clear that knowledge of the expected production (through a better appreciation of weather-based pastoral production forecasts) should be of paramount importance to decision-makers in the New Zealand pastoral industry.

It has been shown that weather conditions, both actual and forecast, can be of considerable value in forecasting New Zealand pastoral production. However, research on the economic value of weather (Maunder 1970, 1978) is only useful as it affects the way decisions are made by managers of weather-sensitive processes. A continuing need exists therefore to assess and present the impact of weather in terms of production figures, costs, or similar measures, which can be used directly by decision-makers — including economists, agriculturists, planners, transportation companies, and politicians. If this is done, the farming community and the nation will reap the benefits.

REFERENCES

Boyer, M. G., 1975 : "The value of weather forecasts in economic planning for the dairy industry", in *Symposium on Meteorology and Food Production*, N.Z. Meteorological Service, Wellington, pp. 213-15.

Maunder, W. J., 1970 : *The Value of the Weather*. Methuen, London, p. 388.

———, 1972 : "National econoclimatic models: problems and applications", *N.Z. Meteorological Service Technical Note 208.*

———, 1974 : The prediction of monthly dairy production in New Zealand through the use of weighted indices of water deficit, *N.Z. Meteorological Service Technical Note 277.*

———, 1977a : "Weather and operational decision-making: the challenge", *Quarterly Predictions,* September, 1977, N.Z. Institute of Economic Research.

———, 1977b : "Weather and climate as factors in forecasting national dairy production", in *Management of Dynamic Systems in New Zealand Agriculture,* D.S.I.R. Information Series 129, pp. 101-26.

———, 1977c : "Climatic constraints to agricultural production, Prediction and planning in the New Zealand setting", *N.Z. Agricultural Science*, 11, pp. 110-19.

———, 1977d : *The Economic, Social and Political Impact of Climatic Variations on Agricultural Production.* Paper presented at the WMO/FAO Technical Conference on the Potential Benefits of Agricultural Meteorology, Rome, October, 1977.

Taylor, N. W., 1977a : Short-term forecasting in the meat industry, in *Management of Dynamic Systems in New Zealand Agriculture,* D.S.I.R. Information Series 129, pp. 79-87.

———, 1977b : "The trend in farm costs and their effect on farm output", *N.Z. Meast and Wool Boards' Economic Service Paper 1775.*

———, 1978 : "The value of incentives and subsidies for increasing livestock production", *N.Z. Meat and Wool Boards' Economic Service Paper 1778.*

Thomson, R. T. and Taylor, N. W., 1975 : "The application of meteorological data to supply forecasting in the meat and wool industry", in *Symposium on Meteorology and Food Production,* N.Z. Meteorological Service, Wellington, pp. 217-29.

Troughton, J. H., 1977 : Preface to Proceedings of a Symposium on the *Management of Dynamic Systems in New Zealand Agriculture,* D.S.I.R. Information Series 129, pp. 5-6.

Walsh, M. J., 1978 : "Notes on farm income production and pastoral exports", *Reserve Bank of N.Z. Publication.*

Ward, F. L., 1978 : "Growth and development of the N.Z. meat and wool industry", *N.Z. Meat and Wool Boards' Economic Service Paper 1786.*

10

THE INDIAN SUMMER MONSOON, ITS ECONOMIC ASPECTS, VAGARIES AND REMEDIAL MEASURES

D. A. MOOLEY

THE MONSOON of Asia is a spectacular phenomenon without any parallel elsewhere in the world. Pedelaborde (1958) has described it as "breathing in" by Asia during summer, and "breathing out" during winter. The word monsoon has its origin in the Arabic word "mawsim" which was used by the sailors for the seasonal alternating winds over the Arabian Sea. Over most parts of South Asia, the surface wind is from sea to land during the summer and from land to sea during the winter. The former is known as the summer monsoon and the latter as the winter monsoon. The summer monsoon, coming as it does from the ocean areas, is associated with large amounts of water vapour and gives copious rain over most parts of South Asia. The winter monsoon, on the contrary, comes from dry-land areas and has got little moisture associated with it, and hence generally gives little rainfall. The summer monsoon is, therefore, of economic importance to most countries in South Asia; 42 per cent of the world population lives in the Asian monsoon area and depends for its livelihood on the summer monsoon. India with a population of about 15 per cent of the world population is located in the central part of South Asia and is predominantly within the regimes of the summer and winter monsoons. The economy of India is vitally linked with the summer monsoon which is also called the south-west monsoon. In fact, the Indian Budget has been often mentioned as a Gamble in the Monsoon.

In this article it is proposed to consider some important features of the Indian summer monsoon and of the rainfall contributed by it, economic aspects of the monsoon, abnormalities in the behaviour of the monsoon over a long period of record, suitable model for the occurrence of the abnormalities, the consequences of these abnormalities and the remedial measures.

What is Monsoon

Conrad (1937) has given an objective definition of the monsoon. According to him, monsoon is a wind regime in which seasonal wind shift occurs. Trewartha (1954) feels that the term "monsoon" may be used for any wind system in which there is a reversal of the prevailing wind direction from winter to summer. Flohn (1960) considers monsoon as a wind system having predominant air currents with a marked seasonal shift resulting from differential heating of continents and oceans during the summer and winter seasons. The Indian summer monsoon is the wind from the Indian Ocean, Arabian Sea and the Bay of Bengal to the Indian land area under the influence of a low-pressure area extending from west Pakistan to Burma across the plains of north India. This low-pressure area results from the differential heating of the land and the sea. The monsoon wind is generally from west to south-west over the Peninsula and the central parts of the country, and from east to southeast over north India. The country is normally under the regime of the monsoon wind and the associated phenomena from June to September. However, variations from the normal do occur in individual years. To the sailor, the monsoon wind is more important. But for the whole country, in general, it is the rain produced by the monsoon which is the elixir for a prosperous economy.

Cause of the Summer Monsoon

Sea and land breezes are diurnal phenomena commonly observed at a coastal station. These are basically due to differential response of water and land to radiation. The specific heat of water is 1.0 and that of soil is 0.1 to 0.2. Hence, if the same amount of heat is supplied to 1 gm of water and 1 gm of soil, the soil will register a rise of temperature five times or more of that registered by water. Moreover, radiation can penetrate deeper in water than in soil, and evaporation at water surface leads to some cooling of the water surface. In view of these factors, during the afternoon, the land gets heated much more as compared to the sea, leading to lower pressure over the land and higher pressure over the sea. Hence a breeze from sea to land, i.e. the cool sea breeze, is initiated during the afternoon. Since land cools much more quickly than the sea during the night, the situation is reversed and early in the morning there is a breeze from land to sea, i.e. land breeze.

The summer and winter monsoons can be considered as sea and land breezes on a very large scale. If there is any part of the world where a huge land mass is in juxtaposition to a huge ocean, it is only the longitudinal belt 40°E to 120°E. In this longitudinal belt the continent of Asia extends from near the equator to 70°N and the Indian Ocean which is south of Asia extends to the Antarctic Circle. During the period March 21 to June 22 the solar elevation at mid-day increases continuously and the land mass of southern Asia gets progressively heated. This leads to the development of low pressure over the area extending from Arabia to Bangladesh across north India by the end of May. The Indian Ocean remains relatively much cooler, and the pressure over the ocean is relatively much higher than that over the land mass of south Asia. Hence the wind blows from the ocean to the land around the low pressure area over north India. Thus the differential heating of the land and the sea is the basic cause of the Indian summer monsoon. It may be mentioned that the atmosphere is not directly heated by the solar radiation. The earth's surface, and the latent heat of condensation of the water vapour in the atmosphere heat the atmosphere. The low-pressure area over north India, also referred to as the monsoon trough, is well established by the end of June. Hereafter, the word monsoon will refer to summer monsoon.

Onset and Withdrawal of the Monsoon

According to the normal dates of onset of the monsoon as worked out by the India Meteorological Department (1978) on the basis of the normal five-day rainfall at observatories in India, the monsoon normally sets in over Trivandrum towards the end of May and by about the first week of July, i.e. during the course of about a month, advances over the whole country. In individual years, the advance of the monsoon shows variations. Ramdas et al. (1954) have worked out the dates of onset of the monsoon over Cochin and Bombay for the period 1891 to 1950. Bhuller (1952) has obtained the dates of establishment of the monsoon over Delhi during the period 1901-50. The lower quartile, the median and the upper quartiles of the distribution of the date of establishment of the monsoon, and the extreme dates together with years for Cochin, Bombay and Delhi are given in Table 10.1.

TABLE 10.1: Onset of the monsoon

	Cochin	Bombay	Delhi
Lower Quartile	May 24	June 4	June 25
Median date	May 29	June 8	July 2
Upper Quartile	June 3	June 12	July 9
Earliest date with year	May 7, 1918	May 21, 1943	June 18, 1925
Latest date with year	June 10, 1935	June 28, 1940	July 20, 1907

The extreme variations in the onset date at these three places is about a month.

The time interval between the onset of the monsoon at Cochin and at Delhi gives an idea of the time taken by monsoon to advance over most of the country after its onset over the southernmost part. The extreme values of this interval are 13 and 61 days. The median, the lower quartile and the upper quartile of the distribution of this interval are 34, 28, 44 days respectively.

The withdrawal of the monsoon from north-west India commences by about the first week of September and the process of withdrawal from most of the country takes normally about a month or so. The withdrawal in some years takes place much earlier than normal. Whereas in some years it is much more delayed than the normal.

The Role of the Mountain System

The mountains affect the air-flow. Higher the height of the mountains, higher is the level up to which the air-flow is affected. According to a diagnostic study of Godbole (1973), in the absence of the Himalayas, the strong westerly wind in the lower troposphere and the strong easterly wind in the upper atmosphere, which are the characteristic features of the monsoon circulation, would not exist. The westerlies and the easterlies would be much weaker, if the Himalayas did not exist.

The effect of the mountains on rainfall distribution is quite well known. Rainfall is more on the windward side and much less on the leeward side. The higher the height of the mountain range, the greater is the contrast in the rainfall on the windward and the leeward sides.

If the Himalayas had not existed, the moist monsoon winds would have gone to higher latitudes and the Indian monsoon rainfall would have been much lower as compared to the monsoon rain that occurs over India.

The mountains thus play a very important role in the monsoon circulation and monsoon rainfall over India.

Statistical Parameters of Monsoon Rainfall

In some years, the monsoon sets in much earlier than normal and withdraws much later than normal. In some parts of the country, the rains during May and October are useful for some of the preliminary agricultural operation for Kharif/Rabi, i.e. summer/winter crops. Whatever monsoon rainfall occurs during October provides soil moisture for Rabi crops. In view of these considerations, the rainfall during the period May to October can be taken as monsoon rainfall. In this study May to October rainfall has been taken as monsoon rainfall.

Normal concentration of rainfall during the monsoon season is shown in Fig. 10.1 for the various meteorological sub-divisions. It is the normal contribution of the monsoon season to the normal annual rainfall of each of the sub-divisions and is expressed as percentage of the annual rainfall. It is based on the data of about 3,000 rain-gauge stations for 1901-60. A large portion of the country gets more than 90 per cent of the annual rainfall during the monsoon. This portion covers about 68 per cent of the country's area and has about the same percentage of India's population. It is this heavy concentration of rainfall over a large part of India during the monsoon season that makes country critically dependent for its prosperity on the cooperative behaviour of the monsoon.

The mean monsoon rainfall (in cm) and the coefficient of variation (standard deviation mean expressed as percentage) are shown in Fig. 10.2 for each of the sub-divisions of the country. It can be seen from Fig. 10.2 that over north India rainfall decreases as we go from east to west, west Rajasthan having the lowest mean rainfall of about 28 cm. Over the Peninsula, rainfall is high over the west coast, decreases rather substantially to the east of the Western Ghats up to some distance and thereafter increases eastwards. Some parts of the Deccan Plateau, like north interior Karnataka, Rayalaseema and Marathwada get low rainfall, 55—75 cm. About 70 per cent of the country receives monsoon rainfall of 100 cm or less. The coefficient of variation is high over northwest India and central parts of Peninsular India, which are regions of low rainfall. It is low over regions of high rainfall, viz. north-east India and the west coast of India.

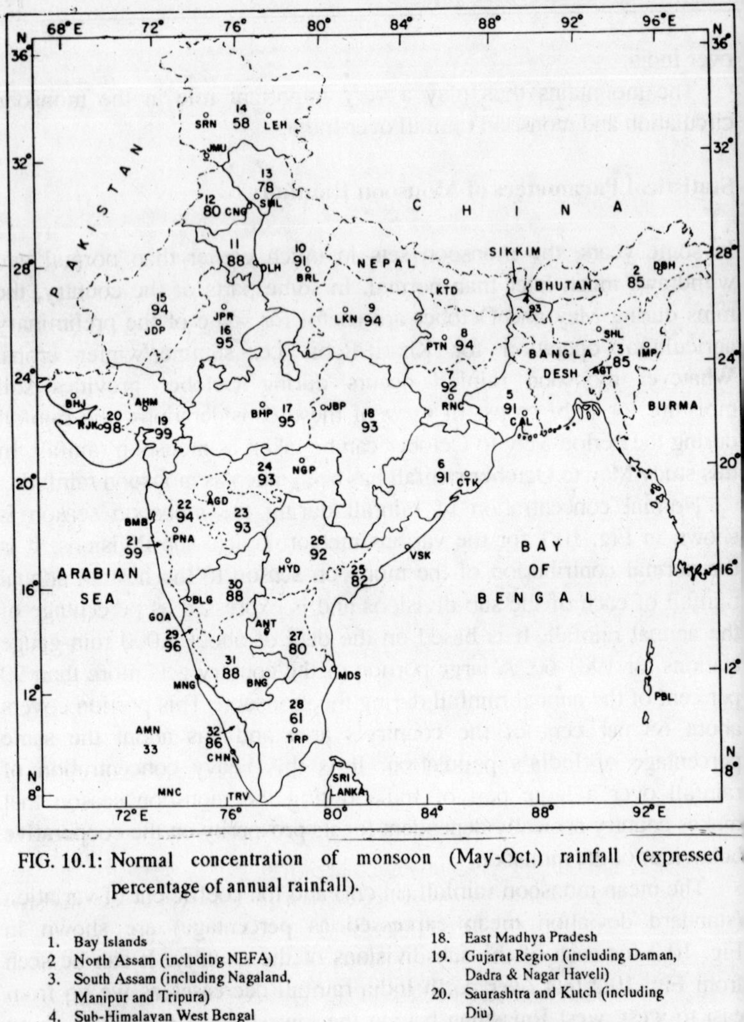

FIG. 10.1: Normal concentration of monsoon (May-Oct.) rainfall (expressed percentage of annual rainfall).

1. Bay Islands
2. North Assam (including NEFA)
3. South Assam (including Nagaland, Manipur and Tripura)
4. Sub-Himalayan West Bengal
5. Gangetic Bengal
6. Orissa
7. Bihar Plateau
8. Bihar Plains
9. East Uttar Pradesh
10. West Uttar Pradesh
11. Haryana (including Chandigarh and Delhi)
12. Punjab
13. Himachal Pradesh
14. Jammu and Kashmir
15. West Rajasthan
16. East Rajasthan
17. West Madhya Pradesh
18. East Madhya Pradesh
19. Gujarat Region (including Daman, Dadra & Nagar Haveli)
20. Saurashtra and Kutch (including Diu)
21. Konkan (including Goa)
22. Madhya Maharashtra
23. Marathwada
24. Vidarbha
25. Coastal Andhra Pradesh
26. Telangana
27. Rayalaseema
28. Tamil Nadu (including Pondicherry)
29. Coastal Karnataka
30. North Karnataka
31. South Karnataka
32. Kerala
33. Arabian Sea Islands

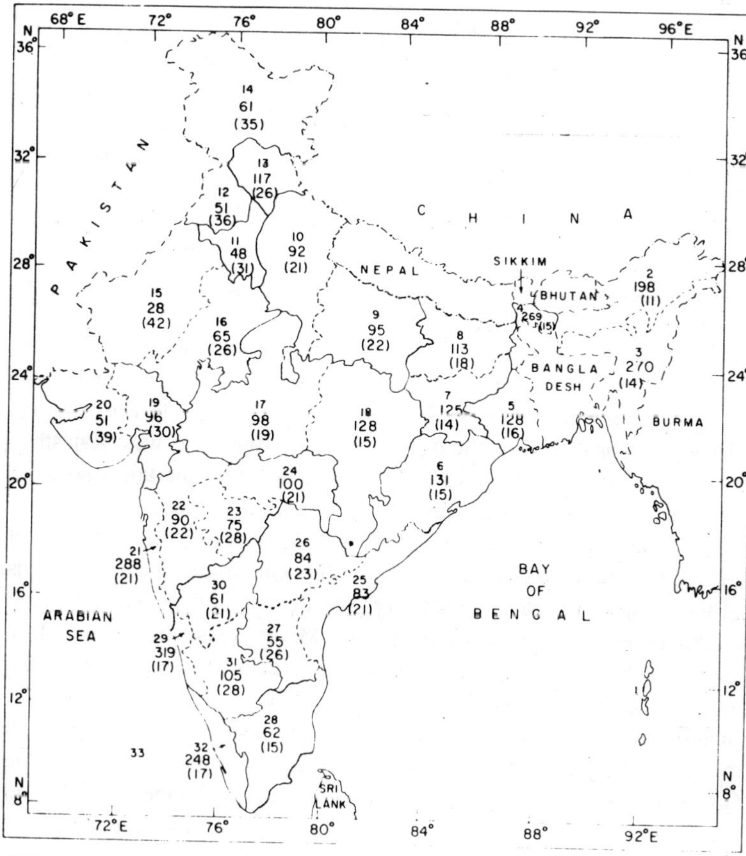

FIG. 10.2: Mean rainfall during May-Oct. and coefficient of variation, mean is indicated below sub-division No. and co-efficient of variation is indicated in brackets below the mean. Rainfall (cm) Coeff Var. (%).

Monsoon Activity and its Variation During the Season

After the monsoon has advanced over the country, its activity is not uniform during the season, but undergoes variations. The synoptic factors which are responsible for the variation in the monsoon activity have been brought out by Mooley (1975). The main synoptic system is the monsoon trough extending from west Pakistan to Bangladesh across north India. Variations in the location and intensity of the monsoon trough due to the perturbations caused by the westward-moving lows, depressions and tropical storms, and the eastward-moving western disturbances across northernmost parts of India result in the variations in the activity of the monsoon. In addition, the variations in the intensity and location of the upper-air anti-cyclone over Tibet affect the activity of the monsoon. From the studies by Mooley (1976, 1978) of the patterns of the monthly monsoon activities, it is seen that for the country as a whole July is the month of maximum monsoon activity and that over the west coast belt between 9°N and 18°N and the adjoining Western Ghats there is a well-marked discontinuity in the monthly activity of the monsoon, i.e. a well-marked decrease in rainfall, from July to August.

We shall now consider the behaviour of the main components of the monsoon system, viz. the monsoon trough, lows, depressions, cyclonic storms, the western disturbances and the Tibetan anti-cyclone.

The monsoon trough is established generally by the end of June. During July and August, the trough normally runs from west Rajasthan to Bangladesh. Convergence and vertical uplift of air are associated with the monsoon trough, particularly in the portion south of the axis of the trough and this results in good rainfall over the central parts of the country. Relatively much less rainfall occurs to the north of the axis of the trough. Rainfall activity over the Peninsula depends in general on the intensity of the monsoon trough. More intense the trough, more intense is the monsoon circulation and this leads to more rain over the west coast, central parts of the country and most of the Peninsula. Reverse happens when the monsoon trough is weak. When the trough moves further south than normal, rainfall is much less over the Gangetic plains. When it moves further north, the Gangetic plains and the sub-Himalayan region get more rain. Sometimes the monsoon trough moves over to the sub-Himalayan region and when this happens rainfall over most of the country decreases markedly while that over the sub-Himalayan region increases considerably, leading to flooding in the rivers originating in the Himalayas and in the States of Assam, Bihar and East U.P. There is also an increase in thunder-shower activity over south Peninsula. This situation is referred to as "break" in the monsoon. These "breaks" have been studied by Koteswaram (1950), Malurkar (1950), Raman (1955), Parthasarathy (1960), Ramaswamy (1962) and Ramamurthy (1969). The breaks in July and August during the period 1888-1967 have been examined in detail by Ramamurthy (1969). During the period 1888-1967, there were 306 days of "break" in the monsoon during July and 356 days in August. This shows that August is more prone to the "break" in monsoon. During July, 48 per cent of the spells of "break" in monsoon are of length exceeding four days, and 14 per cent, of length exceeding eight days. The corresponding percentages for August are 55 and 17. The longest spell in July was of 17 days, July 7–23, 1918, and that in August was of 21 days, July 29—August 18, 1906. The meteorological sub-divisions which suffer most during the spell of "break" are, Saurashtra and Kutch, Gujarat, West Madhya Pradesh, Vidarbha, West Rajasthan and East Rajasthan, the rainfall in these sub-divisions during these spells being less than 25 per cent of the normal. The sub-divisions which benefit most are south Assam, Sub-Himalayan West Bengal and Tamil Nadu, the rainfall in these sub-divisions during these spells being about twice the normal. It is generally observed that a "break" in the monsoon occurs when a low moves from west to east across the northernmost parts of India and simultaneously there is absence of monsoon depressions over the north Bay or over north India. The "break" is normally terminated and there is revival of monsoon over the country when a depression forms over the north Bay and moves west-northwest.

Most of the lows develop over the north Bay. A few of them, however, enter the north Bay after travelling from the China Sea across Vietnam, Cambodia, Thailand and Burma. These lows move within the monsoon trough in a west-northwesterly direction, travelling across Orissa, Madhya Pradesh and Rajasthan, whereafter, they merge with the semi-permanent lows over west Pakistan. Some of them merge with the monsoon trough even before reaching Rajasthan. These lows generally give good rainfall to the south of their tracks. The passage of these lows results in upward transport of moisture and heat and thus they maintain the activity of the monsoon trough.

Some of the lows over the north Bay of Bengal intensify into depressions which generally move in a direction between west and northwest, across the central parts of the country. Some of these travel up to Madhya Pradesh and weaken and merge with the monsoon trough. Some move up to west Rajasthan and then weaken and merge with the seasonal lows over Pakistan. During the month of September, some of the depressions, when they come to east Rajasthan, recurve towards north and north-east and strike the western Himalayas. About 90 per cent of the depressions which move across the country during July and August have their tracks within the belt formed by the line joining Calcutta-Allahabad-Delhi and the line joining Gopalpur-Nagpur-Ahmedabad. Table 10.2 from the study by Mooley (1973) shows the number of Indian monsoon depressions that crossed the different longitudes during the period 1891-1960.

TABLE 10.2: Number of monsoon depressions crossing different longitudes (1891-1960)

Month	90°E	85°E	80°E	75°E	70°E
July	12	96	59	25	12
August	15	103	67	27	2
Total	27	199	126	52	14

It will be seen in this table that of the depressions which cross the Orissa-Bengal coast (85°E), 60 per cent reach the Longitude 80°E, 25 per cent reach 75°E and only 7 per cent reach 70°E. The rate of movement of these depressions varies widely, between 100 and 600 km. per day. These depressions give good rainfall to the south of their tracks. The slow-moving ones give very good rainfall over the central parts. Monsoon depressions contribute 10—15 per cent of the normal rainfall. Heaviest rainfall associated with a monsoon depression occurs in the south-west quadrant of the depression.

Some of the depressions over the north Bay intensify into cyclonic storms and move towards the Indian coast. Soon after striking the coastal belt, they generally weaken into depressions. These storms give very good rainfall over the coastal belt on crossing the coast. The frequency of tropical storms forming over the north and adjoining central Bay is 0.5 per year for each of the months June and July, and 0.4 per year for each of the months August and September. In the Arabian Sea, the frequency of tropical storms is 0.2 per year during June and is negligible in other months.

The lows and troughs which move from west to east across north Pakistan and the northernmost parts of India are referred to as western disturbances. Their role in the production of weather over India has been brought out by Pisharoty and Desai (1956), Kulkarni (1956), Mooley (1957) and Chakravorty and Basu (1957). Their major role during the monsoon season is to usher in monsoon over Punjab, Haryana and Kashmir towards the end of June, to cause heavy spells of rain over the northern parts of India, to shift the monsoon trough northwards to foot Himalayas resulting in a "break" in the monsoon.

The Tibetan anti-cyclone in the upper troposphere is established by the end of June, and it is well-marked at 14 km. level. The variation in the location and intensity of this anti-cyclone affects the upper tropospheric circulation and ultimately affects the monsoon activity.

Trends and Periodicities in the Monsoon Rainfall of India

Parthasarathy and Mooley (1978) have constructed a summer monsoon (June to September) rainfall series for India as a whole for the period 1866 to 1970 and analysed the same statistically to understand the overall behaviour of the monsoon over the country. On the basis of application of the Swed and Eisenhart's runs test for runs above and below the median, and the Mann-Kendall rank statistic test they find that the series neither shows any significant trend nor any significant oscillation. Power spectrum analysis of the series indicates a weak Quasi Beinnial Oscillation (period 2.3 to 2.8 years). A few random occasions of large deviations of monsoon rainfall are found to occur. However, there is a tendency for the Indian monsoon rainfall to return to the normal and there is no cause for fear of any long-term, decreasing trend in monsoon rainfall. Parthasarathy and Mooley (1978) examined the decade mean and standard deviation during the period 1871 to 1970. The values of decade means and standard deviation obtained by them are given in Table 10.3.

The decade mean progressively increased at the rate of about two centimetres per decade from the decade 1911-20 to 1941-50. However, the increase was not significantly different from the overall mean at five per cent level. After the decade 1941-50, the decade rainfall showed a decrease.

TABLE 10.3: Decade mean values and standard deviation

Standard decade	Average monsoon (June-Sept.) rainfall (cm.) of India	Standard deviation
1871-80	87.07	8.32
1881-90	89.61	2.94
1891-1900	87.01	8.51
1901-10	86.83	7.95
1911-20	86.89	12.01
1921-30	88.90	4.76
1931-40	90.87	5.23
1941-50	93.17	5.90
1951-60	91.40	8.36
1961-70	86.99	9.79

Economic Benefits of the Monsoon

The summer monsoon contributes 80—95 per cent of the annual rain over the country and strong winds in some parts of the country. It provides the water potential which is vital for meeting the country's requirements of water for agriculture, irrigation of lands, generation of hydro-electricity and civic needs, and the wind energy. We shall consider both of these aspects.

Water Potential

On an average India receives rainfall of about 97.5 cm during the period May-to-October. Thus every square metre of land surface gets through rainfall 0.975 cubic metre of water. Since the area of the country is about 32×10^{11} sq. m., the total water provided by the monsoon is 31.2×10^{11} cu. m. Of this about one-third is lost through evaporation and about one-fifth through seepage in the soil. What is left is about 45 per cent, i.e. 14.2×10^{11} cu. m. or 142 million hectare-metres, part of which gets utilised for agriculture in the non-irrigated areas, part is used for irrigation of land through canals, tanks and wells, part is used for generation of hydro-electricity, part is used for industry and civic needs, and the rest goes waste into the seas through the Indian rivers. According to Varma (1978), the total annual surface water flows in the country are 178 million hectare metres (m.ha.m), of which 70 m.ha.m. (i.e. 40%) are utilisable and 25 m.ha.m. was the approximate utilisation in 1974. The requirement of fresh water during 1974 was 38 m.ha.m. The requirements for the years 2000 and 2025, as estimated by the National Commission on Agriculture are, 75 and 105 m.ha.m. respectively, out of which water use for irrigation would be about 63 and 77 m.ha.m. respectively, and the balance being the requirement of domestic and livestock, industries and thermal power. As the requirements increase, the country has to harness more and more of this monsoon rainwater which has been flowing into the seas, to meet the growing needs of irrigation, hydro-electricity and water for industry and other purposes. As we see, the present upper limit to the utilisable river flows is 70 m.ha.m. Even though with improvements in technology we may be able to raise this limit, it may not be possible to raise it by more than 25 per cent. Hence to make up the deficiency, we must arrange for the most efficient utilisation of water, recycling of used water, desalination of sea water and adoption of processes requiring the maximum use of water.

During the period 1901-20 the area irrigated was 14-18 per cent of the cultivated area, whereas during the period 1921-47 the same was 18-20 per cent. There has been a good increase in this area in the post-independence period. In 1950-51, it was 21 per cent, in 1968-69, 28 per cent, in 1973-74, 30 per cent and in 1974-75, 31 per cent. According to Planning Commission estimates (*Times of India Year Book*, 1977) only 107 million hectares of 169 million hectares of the gross cropped area, i.e. about 63 per cent, can be ultimately irrigated.

The growth of hydro-electric power and the same expressed as percentage of the total power during the period 1958 to 1978 is shown in Fig. 10.3. The figures shown are for the financial years commencing on April 1 of the year indicated and give the installed capacities. The hydro-electric power is about 40 per cent of the total power. In the country, there were 82 major hydro-electric power projects in 1974. The thermal power has to depend on fossil fuels whose reserves are limited. Moreover, thermal power stations add to the atmospheric pollution. The resources for hydro-electric power, viz. monsoon rain-water, are perennial, and the generation of this power does not add to the pollution. However, the challenge posed by the variability of the monsoon has to be faced.

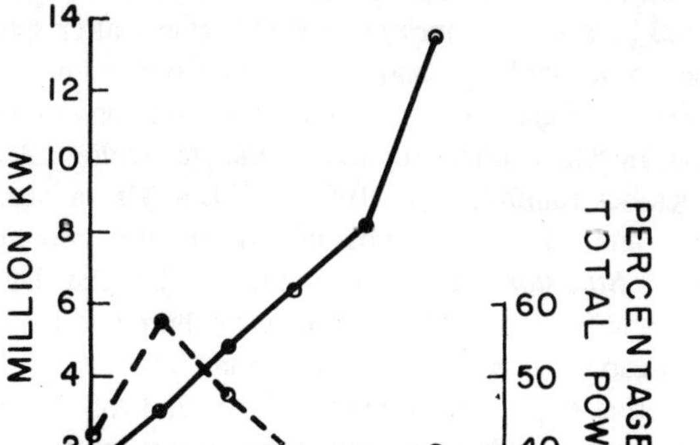

FIG. 10.3: Growth of hydro-electric power in India (1958-78).

The per-capita consumption of power in India is very low. The figures of per-capita consumption for some of the advanced countries in 1968 relative to those for India are given below:

India	1	West Germany	43
U.S.A.	92	France	32
USSR	24	Sweden	91
U.K.	49	Hungary	18

India's per-capita consumption in 1968 was only 71 kilowatt-hours. Therefore, India needs to develop a lot of power for electrifying its villages. A part of these needs can be met by harnessing the monsoon water.

Since India's independence, a systematic effort is being made to increase the food output. The major thrust in this direction is through increase in irrigation, increasing use of fertilisers, high-yielding varieties of seeds, treatment and control of crop diseases. This has yielded good results. It cannot, however, be denied that the overriding factor is the cooperative behaviour of the monsoon. Figures 10.4 and 10.5 bear this out. Figure 10.4 gives the annual foodgrains production (cereals and pulses) in tonnes per 1,000 hectares after subtracting the trend due to technology, and May-to-October rainfall, during the period 1951-76. Figure 10.5 gives the annual rice production in tonnes per 1,000 hectares after removing the technological trend, and May-to-October rainfall, from 1951 to 1976. The annual figures of foodgrains and rice were obtained from the monthly journal, *Agricultural Situation in India* (published by the Directorate of Economics and Statistics, Ministry of Agriculture and Irrigation, Govt. of India). It can be seen from Figs. 10.4 and 10.5 that the monsoon was non-cooperative during the years 1951, 1957, 1965, 1966, 1972 and 1974, and that the production of the total foodgrains as well as that of rice suffered very badly and the country had to import foodgrains substantially. During the years 1953, 1956, 1958, 1970 and 1975 when the monsoon was very cooperative, the production of total foodgrains, and of rice was good.

The correlation coefficient between the annual rice production (after removal of trend due to advancement of technology) and the monsoon rainfall is 0.62 which is significant at one per cent level, whereas that between total foodgrains (with technology trend removed) and monsoon rainfall is 0.40 which is significant at five per cent level.

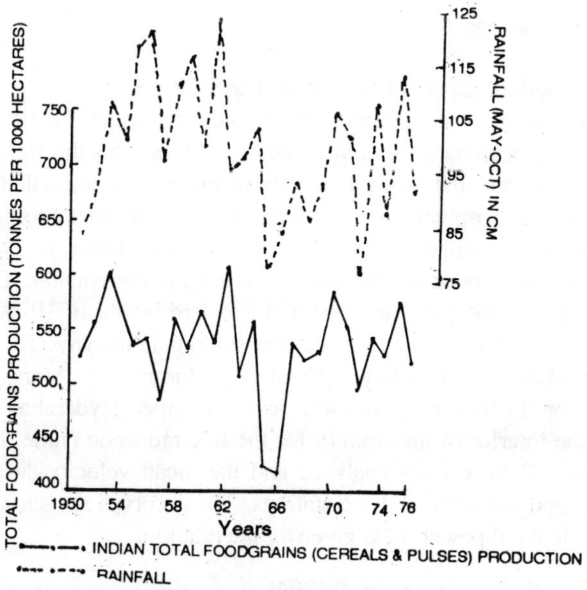

FIG. 10.4: Total foodgrain production in India and monsoon rainfall.

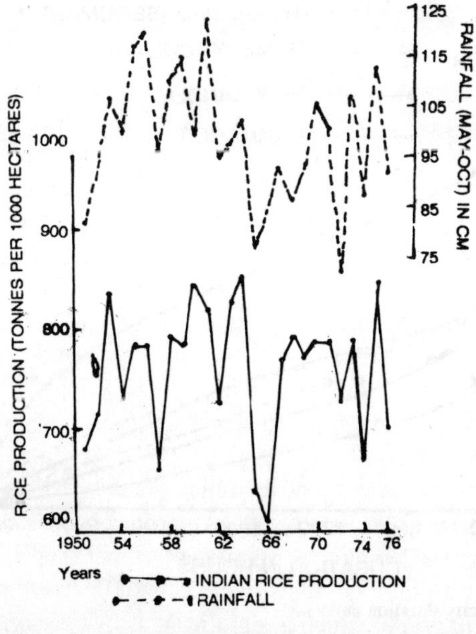

FIG. 10.5: Annual rice production in India and monsoon rainfall.

Power of the Monsoon Wind

The other important aspect of the summer monsoon is the strong wind in the belt of the country dominated by the westerly winds. So far little wind power has been harnessed. West coast of India faces the full blast of the monsoon and the westerly winds in the lowest one-kilometre layer of the atmosphere are strong. The India Meteorological Department has installed self-recording wind instruments to record wind direction and speed at a number of stations in the country. These instruments are exposed to the wind at a standard height of 10 metres above the ground. Hourly wind data of six stations in the westerly belt, viz. New Kandla, Bombay (Colaba), Marmagao, Mangalore (Penambur) on the west coast of India, and Begumpet (Hyderabad) and Bhopal in the interior of the country for the five monsoon (June-Sept.) seasons 1969-73 have been analysed and the mean velocity-duration curves prepared for each of these stations. These curves are shown in Fig. 10.6. The wind power, P, is given by the relation,

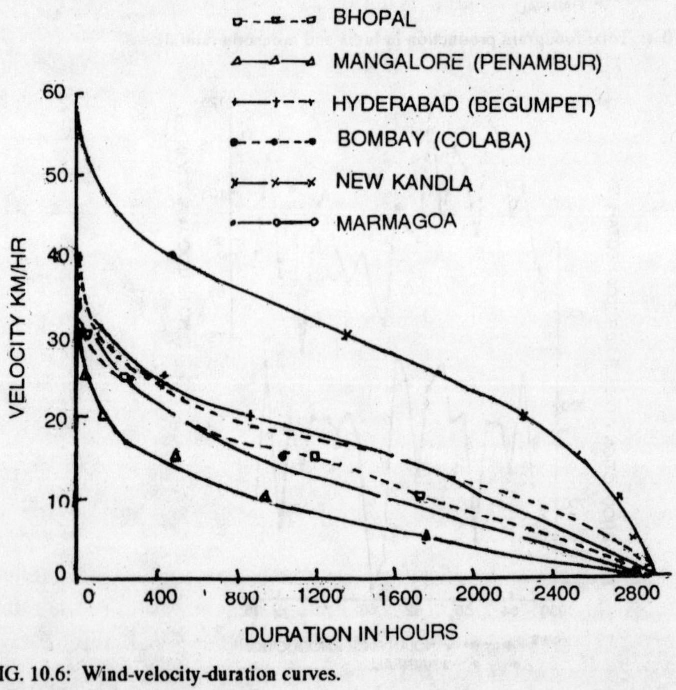

FIG. 10.6: Wind-velocity-duration curves.

$$P = {}^1\!/_2\, \rho A V^3$$

where ρ is air density and A is the area through which wind passes normally and V is the wind velocity. Using the mean value of 1,100 gm/cu.m. of ρ for monsoon months as given by Mooley (1951) and taking A = 100 sq.m., the average power P in kilo-watts is given by

$$P = 0.00117\ V^3$$

where V is velocity in km/hour, and bar denotes mean.

Using this relation and the mean velocity duration curves the mean monsoon (June-Sept.) wind power-duration curves have been obtained. These are given in Fig. 10.7. It can be seen that the monsoon wind at New Kandla shows a good amount of power. The wind power which could be extracted by an ideal aero-motor is about 59 per cent (Golding and Harris, 1976). In actual practice, a windmill may be able to extract only up to 40 per cent of the wind power. The wind power can be used for generation of electricity, grinding corn and drawing water from wells. The perennial monsoon wind power needs to be developed in India. Once this is developed, it would be a boon to the rural life in India.

Vagaries of the Monsoon and Their Implication

The economy of the country is inextricably linked with the monsoon and as such its prosperity is dependent on the good performance of the monsoon. In a vast country like India, it is natural to expect that some parts may experience flood, some may experience drought and the rest may get good rainfall. These unusual deviations in the monsoon rainfall occurring over relatively small areas strain the country's economy to a small extent only. However, there are occasions when large areas of the country are subjected to the vagaries of the monsoon and in these situations the country's economy is severely strained. We shall now consider the large-scale monsoon failures and the large-scale excess rains.

A sub-division of the country is considered to be affected by monsoon failure if the percentage rainfall deficiency from the normal for the period May-to-October is 20 per cent or more. If the rainfall excess in 20 per cent or more, the sub-division is taken to be affected by excess rains. The two sub-divisions, Jammu and Kashmir, and

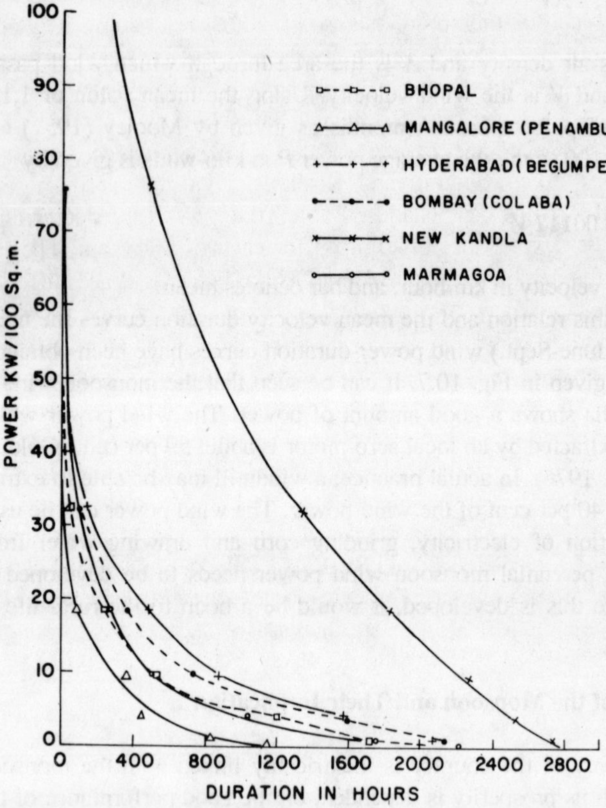

FIG. 10.7: Wind power duration curves.

Himachal Pradesh, have not been considered since these are hilly areas and the available station network would be very inadequate to give satisfactory areal rainfall average. In addition, the two island sub-divisions, the Bay Islands and the Arabian Sea Islands have also not been considered. Thus the area of India considered does not include these four sub-divisions. All the station rainfall data available during the period 1871-1978 have been used to obtain the rainfall series for each of the meteorological sub-divisions. The number of rain-gauge stations whose data were used was about 350 before 1900; 2,000 to 3,000 from 1901 to 1970 and about 350 meteorological observatories after 1970. The normals used for computing the percentage departure from normal are obtained on the basis of data for the period 1871-1978. From the series of percentage departure from normal for each of the sub-divisions, the total area having a rainfall deficiency of 20 per cent or more, as well as the total area having a rainfall excess of 20 per cent or more, have been obtained in each of the years. The total area under each of these two categories is expressed as percentage of the country's area considered, i.e. all sub-divisions except Jammu and Kashmir, Himachal Pradesh, Bay Islands and Arabian Sea Islands. Table 10.4 gives these percentage areas and the sum of these two areas for each of the years. The last column of this table can be considered as the percentage area affected by the monsoon vagaries, i.e. deficient and excess rainfall.

The failure of monsoon is defined to be large-scale when the area covered by monsoon failure is at least 40 per cent of the country's area considered. The years in which the country experienced large-scale failures as seen from Table 10.3 are 1877, 1891, 1899, 1901, 1905, 1911, 1918, 1920, 1941, 1951, 1965, 1966 and 1972. The time interval between successive years of large-scale failure has been tested for randomness by applying both the Swed and Eisenhart's runs test and Mann-Kendall rank statistic test. The tests show that the interval can be taken to be random. The occurrence of large-scale monsoon failures is thus seen to be random in time continuum. The country has to prepare itself to face such large-scale failures which have a frequency of one in about eight years.

TABLE 10.4: Percentage area of the country affected by the vagaries of the monsoon during 1871-1978

Year	Area with rainfall deficiency 20 %	Area with rainfall excess 20 %	Area affected by monsoon vagaries Col. 2 + Co. 3
1	2	3	4
1871	37.0	21.1	58.1
1872	3.3	11.9	15.2
1873	28.5	3.4	31.9
1874	6.9	29.0	35.9
1875	11.3	26.7	38.0
1876	31.8	8.7	40.5
1877	62.7	0	62.7
1878	12.8	43.8	56.6
1879	0	17.1	17.1
1880	21.1	1.8	22.9

(Contd.)

(Contd.)

1	2	3	4
1881	21.3	15.7	37.0
1882	1.6	9.3	10.9
1883	30.2	12.6	42.8
1884	14.3	39.8	54.1
1885	22.0	5.3	27.3
1886	0	28.3	28.3
1887	6.9	22.5	29.4
1888	31.3	8.4	39.7
1889	0	13.5	13.5
1890	21.4	12.5	33.9
1891	41.1	8.2	49.3
1892	2.5	48.2	50.7
1893	6.9	42.2	49.1
1894	3.3	53.5	56.8
1895	29.8	4.5	34.3
1896	28.6	0	28.6
1897	0	5.1	5.1
1898	18.1	11.4	29.5
1899	70.9	6.3	77.2
1900	15.1	13.2	28.3
1901	46.5	0	46.5
1902	26.3	0	26.3
1903	0	21.9	21.9
1904	21.6	0	21.6
1905	48.5	0	48.5
1906	0	3.7	3.7
1907	37.8	0	37.8
1908	11.9	18.0	29.9
1909	12.4	14.0	26.4
1910	0	25.8	25.8
1911	55.6	0	55.6
1912	14.2	5.1	19.3
1913	33.3	12.5	45.8
1914	2.4	18.1	20.5
1915	25.6	13.1	38.7
1916	0	58.6	58.6
1917	0	65.7	65.7
1918	73.1	2.6	75.7
1919	1.8	15.7	17.5
1920	43.0	0	43.0
1921	13.9	13.2	27.1
1922	13.8	12.3	26.1
1923	20.8	1.4	22.2
1924	13.4	11.9	25.3

(Contd.)

(Contd.)

1	2	3	4
1925	32.7	16.1	48.8
1926	4.3	13.6	17.9
1927	10.0	6.7	16.7
1928	13.7	3.4	17.1
1929	20.1	2.6	22.7
1930	5.9	9.8	15.7
1931	0	29.5	29.5
1932	10.4	7.7	18.1
1933	0	62.3	62.3
1934	11.0	13.4	24.4
1935	11.2	0	11.2
1936	21.2	28.2	49.4
1937	6.2	3.0	9.2
1938	14.0	24.7	38.7
1939	32.2	2.5	34.7
1940	15.7	6.7	22.4
1941	61.0	2.5	63.5
1942	9.4	23.4	32.8
1943	3.4	17.5	20.9
1944	8.5	34.8	43.3
1945	2.9	11.9	14.8
1946	8.7	21.5	30.2
1947	1.6	2.6	4.2
1948	16.0	13.2	29.2
1949	8.7	14.5	23.2
1950	3.4	9.2	12.6
1951	46.4	0	46.4
1952	11.7	0	11.7
1953	5.2	27.5	32.7
1954	0	14.2	14.2
1955	0	52.9	52.9
1956	0	59.1	59.1
1957	14.8	6.1	20.9
1958	2.5	31.9	34.4
1959	3.7	47.5	51.2
1960	6.7	4.2	10.9
1961	2.0	60.9	62.9
1962	11.2	18.8	30.0
1963	12.4	7.8	20.2
1964	1.8	18.8	20.6
1965	63.9	0.5	64.4
1966	42.9	6.1	49.0
1967	7.7	3.7	11.4
1968	34.1	0	34.1

(Contd.)

(Contd.)

1	2	3	4
1969	15.0	3.1	18.1
1970	0	24.6	24.6
1971	15.9	22.9	38.8
1972	61.6	4.5	66.1
1973	4.7	47.0	51.7
1974	31.7	7.3	39.0
1975	5.9	41.7	47.6
1976	9.7	13.3	23.0
1977	10.1	28.0	38.1
1978	13.3	33.0	46.3

The rainfall position expressed as percentage departure from normal in the worst years 1877, 1899, 1918 and 1972 is shown in Fig. 10.8. The number of sub-divisions with rainfall deficiencies of ≥40 per cent, ≥50 per cent and ≥60 per cent in each of these four years in Table 10.5.

TABLE 10.5: Number of sub-divisions with rainfall deficiencies

Year	Number of sub-divisions with deficiency		
	≥40%	≥50%	≥60%
1877	6	5	2
1899	13	10	5
1918	12	5	5
1972	4	3	1

It can be seen that the monsoon failure of 1899 was the severe most. The other three years in the order of decreasing severity are 1918, 1877 and 1972. However, since the number of rain-gauge stations was much smaller prior to 1901, as compared to that for the period 1901-70, it is rather difficult to consider undisputed the first rank of the drought of 1899 vis-a-vis the drought of 1918.

Since large-scale monsoon failure has an adverse impact on the economy of the country, it may be advisable to plan and provide suitable funds to mitigate the hardship. For this purpose, it would be necessary to know the probability of getting one, two, three years of large-scale monsoon failures in a five-year period. The probability models which can be tried in this case are the Binomial and the

Poisson. Under certain conditions, there is transition from Binomial to Poisson, and both the models are likely to show good fit. Both the models were fitted to the data and the goodness-of-fit of the models has been tested by Chi-square test. The results are given in Table 10.6. It can be seen from this table that while both these models show a good fit, the fit of the Binomial model is better as brought out by the value of Chi-square. Binomial model could be used to obtain the probabilities of getting 0, 1, 2, 3 years of large-scale monsoon failure in a five-year period.

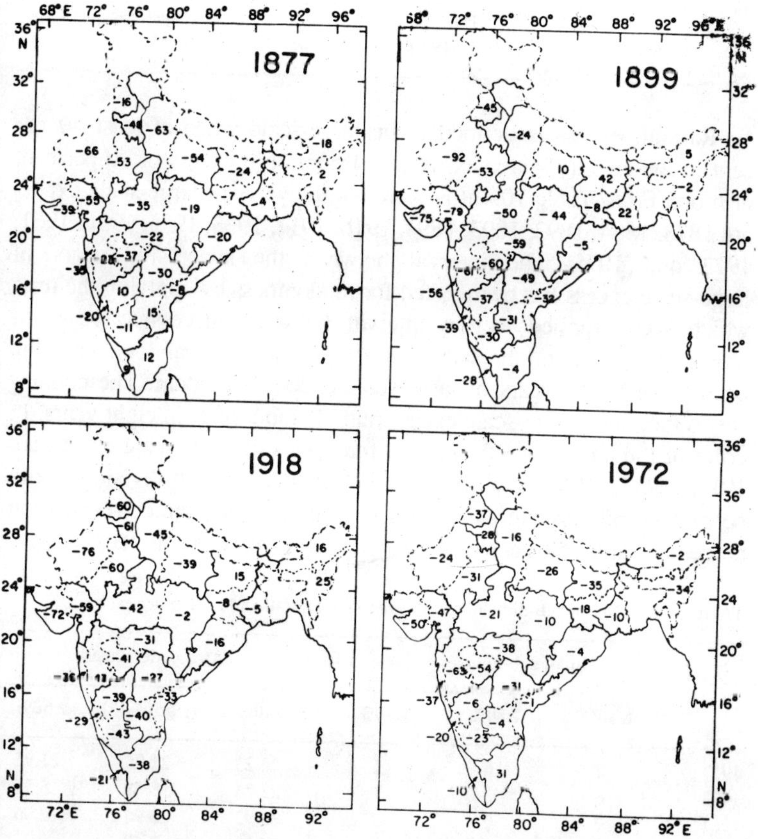

FIG. 10.8: Percentage departure of May-October rainfall from normal during 1877, 1899, 1918 and 1972.

TABLE 10.6: Fit of the Poisson and Binomial distributions to the number of large-scale monsoon failure in five-year periods (1871-1978)

No. of large-scale monsoon failures in 5-year period	Observed frequency	Frequency on Poisson hypothesis	Contribution to Chi-square	Frequency on Binomial hypothesis	Contribution to Chi-square
0	10	11.53	0.203	11.17	0.123
1	9	6.91	0.632	7.41	0.341
2	2	2.08	0.214	2.08	0.073
≥3	0	0.69		0.34	
		Chi-square = 1.049 (d.f.1)		Chi-square = 0.537 (d.f.1)	

Rainfall excess is defined to be large-scale when at least 40 per cent of the country's area experienced rainfall in excess of 20 per cent or more. From Table 10.4 it is seen that the years of large-scale excess are 1878, 1884, 1892, 1893, 1894, 1916, 1917, 1933, 1955, 1956, 1961, 1973 and 1975. The interval between the successive years of large-scale excess has been tested for randomness by applying the tests which were applied to the interval between successive years of monsoon failure, and the same has been found to be random. Thus the occurrence of the years of large-scale excess is random. The country has to face such large-scale excess rainfall about once in eight years. In order of the area covered, the first four years of large-scale excess are 1917, 1933, 1961 and 1956. The number of sub-divisions with percentage rainfall excess of ≥40, ≥50 and ≥60 and the corresponding percentage areas for each of these years are given in Table 10.7.

TABLE 10.7: No. of sub-divisions with excessive rainfall and area under each

Year	No. of sub-divisions with excess			Area under excess		
	≥40%	≥50%	≥60%	≥40%	≥50%	≥60%
1917	7	6	1	26.23	21.95	21.95
1933	8	3	1	23.73	8.51	8.51
1961	7	2	2	25.92	3.96	3.96
1956	6	1	0	24.22	3.67	0

The above figures bring out that the year 1917 was the most severe in respect of rainfall excess. Next in order of severity would be 1933,

1961 and 1956. Since large-scale excess rainfall leads to floods and dislocates the economy of the country, it would be worthwhile to plan and provide suitable funds for flood control schemes. Both the Poisson and Binomial models have been applied to the number of years of large-scale excess in a five-year period. It is found that both the models show a good fit. However, the fit of the Poisson model is better as can be seen from Table 10.8. From the statement of flood damage in India since 1965 as given by *Times of India Year Book* (1977), the total damage to crops, houses and public utilities was about Rs. 500 crores during 1973. The area and population affected were about 16.16 million hectares and 33.09 million, while the cattle and human lives lost were 257,708 and 1,267 respectively.

TABLE 10.8: Fit of the Poisson and Binomial distributions to the number of years of large-scale monsoon rainfall excess in five year periods (1871-1978)

No. of years of large-scale monsoon rain-fall excess in 5-year period	Observed frequency	Frequency on Poisson hypothesis	Contribution to Chi-square	Frequency on Binomial hypothesis	Contribu-tion to Chi-square
0	12	11.53	0.019	11.17	0.062
1	6	6.91	0.120	7.41	0.268
2	2	2.08	0,019	0.08	0.139
≥3	1	0.69		0.34	
		Chi-square = 0.158 (d.f.1)		Chi-square = 0.469 (d.f.1)	

Defining a year of large-scale monsoon vagary as one in which the total area under rainfall deficiency of ≥20 per cent and rainfall excess of ≥20 per cent is 50 per cent or more of the country's area, it is found from Table 10.1 that such years are 1871, 1877, 1878, 1884, 1894, 1899, 1911, 1916, 1917, 1918, 1933, 1941, 1955, 1956, 1959, 1961, 1965, 1972 and 1973. Table 10.9 shows the goodness-of-fit of the Poisson and Binomial models to the number of such years in a five-year period. The fit is good for both. However, the fit of the Poisson model is better.

As a result of the vagaries of the monsoon, the food production suffers rather a serious setback, and there is loss of life particularly in severe floods and severe droughts. Shortfall in rain in catchment areas lowers the hydro-electric potential and leads to power cuts which at times affect the industry very adversely. Severe droughts impose a great strain on the country's economy.

TABLE 10.9: Fit of the Poisson and the Binomial distributions to the years of large-scale monsoon vagaries of the Indian monsoon in five year periods (1871-1978)

No. of years of large-scale vagaries in 5-year period	Observed frequency	Frequency on Poisson hypothesis	Contribution to Chi-square	Frequency on Binomial hypothesis	Contribution to Chi-square
0	8	8.13	0.002	7.33	0.061
1	7	7.71	0.065	8.59	0.295
2	5	3.65	0.137	4.03	0.167
≥3	1	1.51		1.05	
		Chi-square = 0.204 (d.f.1)		Chi-square = 0.523 (d.f.1)	

Measures to Reduce Dependence on the Monsoon

A large part of the available water resources of the country is flowing waste into the sea through the Indian river systems. During the period 1871-1978 there is only one case of large-scale monsoon failure in two successive years. Thus the country is very rarely expected to face the dreadful event of two successive large-scale failures.

One way to exploit the water which is flowing into the sea is to design and execute multipurpose river-valley schemes, increase the area under canal irrigation, tank irrigation and well irrigation. This area which at present is about 30 per cent of the cultivated area may be increased to about 50 per cent through planned and sustained efforts.

The technological advances in the field of agriculture, viz. high-yielding varieties, fertilisers, treatment for crop-diseases, arranging one more crop during the year, etc., may be employed to the maximum in the irrigated area to maximise food output.

Even with the best of efforts, more than 52 per cent of the cropped area cannot be brought under irrigation. Moreover, pulses, oilseeds, cotton and cattle feed are grown mostly on dry lands. In view of this, dry-farming technology should be developed to its maximum and the same should be applied at the earliest. Whatever methods are available at present for increasing the production, viz. using new and drought-resistant varieties, adopting improved package of practices, etc., may be employed. It is estimated that 36 million holdings, constituting 54 per cent of all land holdings in semi-arid rain-fed regions, are smaller than one hectare, and each tiny holding comprises three to four fragments. Hence it is necessary that the Government and its concerned departments and agencies give encouragement and facilities to the land holders to apply properly the useful results of dry-farming research to enhance production in dry lands.

In each State, the drought-prone areas may be identified. Such areas may be given special attention. In this direction, the Government of India has taken since 1970-71 several measures for the development of drought-prone areas under its Drought Prone Areas Programme (DPAP) (*Times of India Year Book*, 1977). These are: (a) Development of irrigation, (b) Formulation of schemes specially designed to benefit these areas, (c) Agricultural research oriented to problems of low rainfall areas, (d) Scarcity relief which also helps in creating additional production potential and provision of services like drinking water supply, etc., and (e) Special programmes like Emergency Agricultural Production Programme. The main emphasis of the DPAP is in the direction of restoration of a proper ecological balance in the selected areas, and covers: (i) Development and management of irrigation, (ii) Soil and moisture conservation and afforestation, (iii) Restructuring of cropping patterns and pasture development, (iv) Livestock development, (v) Popularisation of dry farming technology, and development of small and marginal farmers and agricultural labour. The total amount allocated for this programme in the Fifth Five Year Plan (1974–79) was Rs. 1.87 crores.

To meet the hardship caused by the natural calamities, like droughts, floods and pest attacks, and the consequent loss of the major portion of the crop, the Government of India has recently finalised a pilot crop insurance scheme, with the State Government as the co-insurer (*Times of India, Indian Express*, 1979). The scheme will be for homogeneous agroclimatic blocks with common premia rates and indemnity. The State Government as co-insurer would share claims as well as premia to the extent of 25 per cent. Premium indemnity table has already been prepared for 12 States and one Union Territory. The scheme will be operated by the General Insurance Corporation of India.

Conclusion

The summer monsoon of India is a seasonal wind regime caused by the differential heating of land and sea. The monsoon wind blows from sea to land under the influence of the low pressure over Pakistan, north India and Bangladesh. It undergoes variations in its time of onset and withdrawal and in its activity. Variations in the intensity and location of the monsoon trough caused by the westward-moving perturbations, like lows, depressions and cyclonic storms, and the eastward moving perturbations, like western disturbances moving across northernmost parts of India, result in variations of monsoon rainfall. The Indian mountain system plays a very important role in the genesis of the monsoon circulation and monsoon rainfall distribution over India. Monsoon (May-Oct.) rainfall constitutes more than 90 per cent of annual rain over 68 per cent of the country's area. This situation makes the country suffer very badly in situations of monsoon failure. Monsoon rainfall variability is high over regions of low rainfall.

Amongst the economic benefits of the monsoon which can be further and more intensively exploited for the country's development are rainwater for increasing irrigation and hydro-electric potential and wind power for helping the rural people. In the record of the Indian monsoon for the past 150 years, 1899 and 1918 are the worst years of monsoon failure, and 1917 and 1933, the worst years of excessive monsoon rainfall. While there are a few occasions of large deviations of monsoon rainfall from normal and the consequent suffering by the people there is a tendency for the rainfall to return to normal and there is no cause for fear of any long-term decreasing trend in the Indian monsoon rainfall.

A severe and large-scale monsoon failure upsets the country's economy. The occurrence of either large-scale monsoon failure or large-scale monsoon rainfall excess over India is found to be random. Poisson probability model is found to fit well the number of years of large-scale monsoon failure/large-scale monsoon rainfall excess in a five-year period. The probabilities based on this model could be taken into account while planning for funds to fight against these natural calamities.

REFERENCES

Bhuller, G. S., 1952 : "Onset of monsoon in Delhi, *Ind. J. Met Geophys*, Vol. 3, No. 1, pp. 25-30.

Chakravorty, K. C. and Basu S. C., 1957 : "The influence of Western Disturbances on the weather over north-east India in monsoon months", *Ind. Journ. Met. & Geophys*, Vol. VIII, No. 3, pp. 216-65.

Conrad, V., 1937 : *Meteorologische Zeitschrift*, 54, pp. 313-31.

Flohn, H., 1960 : "Monsoon winds and General Circulation", *Symposium on the Monsoons of the World*, New Delhi, February 1958, pp. 65-74.

Godbole, R. V., 1973 : "Numerical simulation of the Indian Summer Monsoon", *Ind. J. Met. Geophys*, Vol. 24, No. 1, pp. 1-14.

Golding, E. W., 1976 : *The Generation of Electricity by Wind Power*, E & F.N. Spon Ltd., London, p. 191.

Indian Meteorological Department, 1978 : *Agro Climatic Atlas of India*, pp. 2-3, Pune.

Indian Express, 1979 : Crop Insurance Scheme, Newspaper of April 17, 1979.

Koteswaram, P., 1950 : "Upper air lows in low latitudes in the Indian area during southwest monsoon season and 'breaks' in the monsoon", *Ind. J. Met. Geophys*, Vol. I, No. 2, pp. 162-64.

Kulkarni, S. B., 1956 : "A study of Western disturbances and 'breaks', in the southwest monsoon", *Ind. Journ. Met. and Geophys*, Vol. VII, No. 1, p. 5.

Malukar, S. L., 1950 : "Notes on analysis of weather of India and neighbourhood", *Memoirs of the I. Met. D.*, XXVIII, Part IV, pp. 139-215.

Ministry of Agriculture and Irrigation, Govt. of India, 1958-76, *Agricultural Situation in India*, Directorate of Economics and Statistics.

Mooley, D. A., 1951 : "Normal density distribution in the atmosphere", *Ind. J. Met. & Geophys.*, Vol. 2, No. 2, pp. 127-37.

——— 1957 : "The Role of Western Disturbances in the Production of Weather over India during different seasons", *Ind. J. Met. Geophys*, Vol. 8, No. 3, July 1957.

——— 1973 : "Some aspects of Indian monsoon depressions and associated rainfall", *Mon. Wea. Review*, USA, Vol. 101, No. 3, March 1973.

——— 1975 : "Climatology of the Asian Summer Monsoon Rainfall — controls and concentration", *Geographical Review of India*, Vol. 37, No. 1, March 1975.

——— 1976 : "Climatology of the Asian Summer Monsoon Rainfall — (2) Patterns of monthly monsoon activity and their geographical distribution", *Geographical Review of India*, Vol. 38, No. 1, March 1976.

—— 1978 : "Major climatological discontinuities in the monthly monsoon activity in the neighbourhood of the Western Ghats", *Ind. J. Met. Hydrology & Geophys*, Vol. 29, No. 3, July 1978.

Parthasarathy, B. and Mooley, D. A., 1978 : "Some features of a long homogeneous series of Indian summer monsoon, rainfall", *Mon. Wea. Review*, Vol. 104, No. 6, June 1978.

Parthasarathy, K., 1960 : "Some aspects of rainfall in India during the Southwest monsoon season", *Proc. Symp. on Monsoons of the World*, India Met. Dept., pp. 185-94.

Pedelaborde, P., 1958 : *Les Moussons*, published in France by Armand Colin, English Translation, *The Monsoon* published by Methuen and Col. Ltd., 1963 and 1974, p. 8.

Pisharoty, P. R. and Desai, B. N., 1956 : "Western disturbances and Indian weather", *Ind. Jour. Met. and Geophys.*, Vol. VII, pp. 333-38.

Ramamurthy, K., 1969 : "Some aspects of the 'break' in the Indian Southwest monsoon during July and August", *Forecasting Manual*, Part IV, The Deputy Director General of Observatories (Forecasting), Poona-5, Ind. Met. Department.

Raman, C. R. V., 1955 : "Breaks in Indian Southwest monsoon and typhoons in the southwest Pacific", *Current Science*, XXIV, pp. 219-20.

Ramaswamy, C., 1962 : "Breaks in the Indian summer monsoon as a phenomenon of interaction between the easterly and subtropical westerly jet streams", *Tellus*, Vol. XIV, No. 3, pp. 337-49.

Ramdas, L. A., Jagannathan, P. and Gopal Rao, S., 1954 : "Prediction of the date of establishment of southwest monsoon along the west coast of India", *Ind. J. Met. Geophys*, Vol. 5, No. 4, pp. 305-14.

Times of India, 1977 : *Times of India Directory and Year Book* (Editor, Sham Lal), pp. 46-48 and 73-75.

———, 1979 : Pilot Crop Insurance Scheme Finalised, Newspaper of April 18, 1979.

Trewartha, G. T., 1954 : *An Introduction to Climate*, McGraw-Hill Book Company, Inc., New York, p. 94.

Varma, C. V. J., 1978 : "Efficiency in the use and reuse of water in India", *Jour. of Irrigation and Power*, Vol. 35, No. 1, pp. 73-90.

11

CLIMATIC PATTERN AND AGRO-CLIMATIC REGIONS OF ASSAM

M. M. DAS

CLIMATE IS the principal factor of the physical environment affecting agriculture. The characteristics of soil upon which vegetation and agricultural crops flourish are mainly the products of present and past climates. Cropping pattern evolved in traditional peasant agriculture is the expression of climatic pattern to a great extent.

The climate of Assam would have been similar to the general climate pattern of the tropical monsoon lands, had there been no orographic diversity within the State and on its boundaries. The orographic diversity and local physiographic variations are the strongest factors that have modified the Cwg (Humid Mesothermal — Gangetic type) climate as identified by Koppen, because the weather in Assam may be characterised by the predominance of thunderstorms, winter fog and local mountain valley wind. The development of orographic low, a spectacular but complex thermodynamic phenomenon, is an uncommon feature in a Cwg region. Therefore, the variations in the elements and factors deviate the normal Cwg climate in north-east India and, in fact, these deviations have developed a climatic individuality for Assam. Perhaps, it will be more appropriate to distinguish the climate of Assam as Cwa instead of Cwg.*

In general, the climate of Assam is characterised by coolness and extreme humidity. The climate of the hills is generally mild, but the

* Cw—Where rainfall is highly concentrated in mid-summer and winter is dry. This type of climate prevails in monsoon lands.

a—Hot summer, average temperature of the warmest month over 22°C.

g—Ganges type of annual march of temperature, hottest month comes before the solstice and the summer rainy season.

plain is comparatively much warmer in summer and cool in winter. In the plains, the maximum temperature does not go beyond 32°C in summer and does not fall below 8°C in winter. But in the hills, winter can be rather cold. The extreme humidity makes the climate of Assam very unpleasant.

Seasonal Characteristics

Perhaps it would be advantageous to study the climatic conditions of Assam if the year is divided into four seasons — (i) Pre-Monsoon, (ii) Monsoon, (iii) Retreating Monsoon, and (iv) Dry Winter. This cycle of climatic seasons has a great bearing on the agricultural cycle of the State.

The pre-monsoon starts by early March alongwith the gradual rise in temperature coupled with vanishing fog and occasional thunder showers. This season characterised by cool and pleasant mornings, hot and irritating afternoons, occasional duststorms and thunder showers, continues up to the end of April. Very strong winds causing duststorms and thereby making the sky hazy in the month of March is a peculiar climatic phenomenon in the lower Brahmaputra valley reaching up to Tezpur eastwards. Assam gets a good amount of rain falling during this season from the "Norwesters" which makes the climate cool even during spring. In contrast, there is very little spring rainfall from March to May in the rest of northern India. The occasional rain of this season is highly significant for the cultivation of jute, Ahu rice and for the budding of tea plants which are grown abundantly in Upper Assam.

The monsoon breaks in Assam by the middle of May and continues up to September. This season is characterised mainly by cloudy weather, higher humidity, heavy rainfall, higher temperature and weak variable surface wind. Incessant heavy downpour of rainwater derived from the south-west monsoon keeps the summer temperature lower than it would have been in case there were no continuous rainfall. Rainfall is so frequent that about 18 to 20 days in a month are rainy. The monsoon season is the most important period for agricultural activities of the Assamese peasants, since the cultivation of winter rice (*Sali* rice), the principal cereal crop in Assam, takes place during this period.

By the middle of September, south-west monsoon starts to retreat followed progressively by fair weather and morning fogs towards the end of November. During this season, the number of rainy days as well as intensity of rainfall decrease. The sky becomes clearer and free from any kind of disturbances. The moderate autumn rainfall is beneficial to the sowing of winter crops.

Winter season begins at the end of November and continues up to the end of February. The main characteristics of this season are the cool weather with low temperature around 10°C to 12°C and frequent morning fogs. Sometimes, the sun is not seen before noon. In the hills, temperature falls further down. The pleasant weather of this season is occasionally vitiated by showers and cold spells which lower the temperature. Rainless days of the season help the harvesting of *Sali* rice, but hinder the growth of Rabi crops. Therefore, irrigation is the sine qua non in Assam for successful Rabi crops.

Spatial Variation

There is spatial variation of the climatic characteristics described above within the State of Assam. Tables 11.1, 11.2 and 11.3 make it evident that the weather is not homogeneous throughout the State within a particular season. Relief is the main factor for the variation of distribution of the main climatic elements such as rain, humidity and temperature. The traditional peasant agriculture has been adjusted to this varying nature of climatic elements from season to season as well as from place to place. If there is a fluctuation in the timely distribution of any one of these elements, particularly of rainfall, then the seasonal crop cultivated by the peasant is badly damaged. In this way, the climatic condition is still beyond the control of the poor individual cultivators of Assam.

Table 11.2 reveals that there is little variation of the distribution of relative humidity from place to place, but seasonal variation is indicating comparative dryness and dampness is well marked in all the stations. In general, the climate of the State as a whole is damp at all seasons. Such damp climate is most suitable for cultivation of rice but it is difficult to grow wheat.

Table 11.3 shows that while there is little spatial variation of temperature distribution, the seasonal variation is considerable in all the stations. The monthly differences between the maxima and minima are also much higher in all the places. In general, the summer and winter seasons are well contrasting in respect of temperature condition, thereby facilitating the cultivation of a variety of crops during different seasons of the year. For this reason, seasonal rhythm is seen in the agricultural activities of Assam as a whole. The Assamese peasants are getting the benefits of the tropical monsoon climate in the summer as well as of the temperate dry climate in the winter for growing a wide range of crops.

TABLE 11.1: **Season-wise average rainfall and number of rainy days, 1931—60**

(Rainfall in millimetre)

Station	Pre-Monsoon	Monsoon	Retreating Monsoon	Winter	Total
Dibrugarh	303.4	2145.0	193.0	118.0	2759.4
	(19.9)	(89.4)	(10.5)	(11.9)	(131.7)
Sibsagar	312.8	1930.3	165.6	95.7	2504.3
	(16.9)	(82.1)	(11.1)	(9.7)	(125.8)
Tezpur	204.2	1488.2	155.4	46.8	1894.4
	(15.1)	(76.5)	(8.8)	(3.5)	(103.9)
Lumding	170.8	907.6	154.6	82.1	1315.1
	(10.8)	(53.4)	(8.8)	(3.5)	(76.5)
Silchar	409.5	2498.7	251.0	66.2	3225.4
	(19.1)	(100.9)	(10.4)	(4.8)	(135.0)
Gauhati	179.3	1321.6	101.6	34.7	1637.2
	(12.5)	(64.8)	(6.1)	(3.5)	(80.9)
Dhubri	199.5	2147.2	146.9	31.6	2525.2
	(10.7)	(76.1)	(6.1)	(2.5)	(95.4)

Source: Climatological Tables of Observation in India (1931-60).

Note: Figures in brackets indicate the number of rainy days.

TABLE 11.2: **Season-wise mean humidity recorded at 8.30 hrs. (P.C. relative humidity) 1931-60**

Station	Relative Humidity				
	Pre-Monsoon	Monsoon	Retreating Monsoon	Winter	Total
Dibrugarh	74.0	86.0	80.5	85.0	82.8
Sibsagar	78.0	84.6	86.5	89.0	84.9
Tezpur	69.5	85.0	80.0	81.7	80.8
Lumding	67.5	80.2	84.5	81.3	79.1
Silchar	72.5	84.8	82.0	81.7	81.5
Gauhati	71.5	83.8	85.5	85.0	82.3
Dhubri	70.0	86.0	82.0	81.3	81.5

Source: Same as that of Table 11.1.

TABLE 11.3: Seasonal mean temperature in C°, 1931—60

Month	Dibrugarh		Sibsagar		Tezpur		Lumding		Silchar		Gauhati		Dhubri	
	Max.	Min.	Max.	Min.	Max.	Min.	Max.	Min.	Max.	Min.	Max.	Min.	Max.	Min.
January	22.5	10.2	22.7	9.9	23.7	11.4	24.0	11.0	22.9	11.8	23.9	8.6	25.5	11.7
February	23.6	12.9	23.9	12.8	25.4	13.7	26.3	12.8	25.3	14.2	26.2	10.6	27.1	13.6
March	26.7	15.8	27.3	16.1	29.0	17.1	30.2	16.5	30.1	18.1	30.9	14.7	30.4	17.58
April	28.2	19.2	28.7	19.3	30.3	20.2	31.6	20.3	31.6	21.5	33.0	19.1	31.5	20.8
May	29.2	21.7	29.7	22.4	30.3	22.5	31.0	22.7	29.9	23.0	32.5	22.1	31.2	22.1
June	30.2	23.7	31.5	24.8	31.6	24.7	31.5	24.7	29.8	24.6	32.5	24.2	31.3	24.6
July	30.7	24.5	32.4	25.4	32.1	25.3	32.1	25.8	29.6	25.8	33.0	24.8	31.8	25.1
August	30.9	24.7	32.1	25.4	32.2	25.4	32.2	25.8	30.7	26.0	32.9	24.7	31.2	24.5
September	30.2	23.9	31.5	24.7	31.7	24.7	32.1	25.2	30.7	24.7	32.3	24.1	31.8	24.7
October	29.5	21.0	29.7	21.8	30.2	21.8	30.5	22.0	29.1	23.0	30.2	21.3	31.1	22.7
November	27.1	15.6	26.8	16.0	27.6	16.5	27.7	16.9	26.3	17.8	27.1	15.4	29.2	17.6
December	24.0	11.1	23.8	11.4	24.7	12.4	24.9	15.5	23.7	14.3	24.4	10.6	26.5	13.3
Annual Mean	27.7	18.7	28.3	19.2	29.1	19.6	29.5	19.7	28.3	20.4	29.9	18.3	29.9	19.9

Source: Same as that of Table 11.1.

TABLE 11.4: Temperature in Degrees Centigrade

Station	Maximum	Minimum	Annual Range of temperature
Dibrugarh	22.5	10.2	12.3
(Upper Brahmaputra Valley)	(30.9)	(24.7)	(6.2)
Tezpur	23.7	11.4	12.3
(North Bank Plain)	(32.2)	(25.4)	(7.8)
Lumding	24.0	11.0	13.0
(Central Brahmaputra Valley)	(32.2)	(25.8)	(7.4)
Silchar	22.9	11.8	11.1
(Barak Valley)	(30.7)	(26..0)	(4.7)
Gauhati	23.9	8.6	15.3
(Lower Brahmaputra Valley)	(33.0)	(24.8)	(8.2)

Note: Figures within brackets indicate those of summer season.

The annual range of temperature is higher in the winter season than in the summer season. On the whole it is not very high in comparison to other north Indian Sates of India. Though it is evident from Table 11.4 that there are two distinct seasons, viz. summer and winter, the temperature is favourable for crop husbandry throughout the year, because it is above the conventionally accepted threshold temperature, that is, 6°C (Schimper, 1903) at which active germination and growth both begin to take place with most useful crops. The two best indicators of regional differences in temperature are the length of growing season and accumulated temperatures above the minimum for plant growth. In Assam, the length of growing season and temperature nowhere become the limiting factors to cropping. Assam is blessed with the change of seasons which enables it to grow a variety of crops with or without developed water supply, because a growing season is never limited by cold.

The Climogenetic Supply of Water : Rainfall

Rainfall is the most important weather element for agriculture. Almost the total available supply of water for growing crops in Assam is met by rainfall, other sources of water supply if any, playing an insignificant role. Peasant agriculture in this State may, therefore, be called "a Gamble in Monsoon". Though heavy monsoon rainfall occurs every year in Assam, it varies from year to year and from place to place. The average annual rainfall of the State during the past 26 years

(1951–76) was 2,532 mm. But there is variability of monsoon rainfall which is of considerable importance for agriculture (Table 11.4). Regional variation of rainfall variability is shown in Fig. 11.1 with the help of the index of coefficient of variability. Generally, variability being largest in the driest and least in the wettest regions.

Though Assam is said to fall within the zone of assured rainfall (1,140 mm. and above in a year) in the country, year-to year as well as season-to-season fluctuation is observed. Periodic cycles of flood and drought have been experienced at the interval of three to four years. Agricultural crops in this State are badly affected by such cyclical fluctuations of rainfall. Figure 11.2 shows clearly the negative correlation between the annual average income from agriculture and the areas affected by floods.

Absence of timely rainfall hampers timely operation of agriculture by the peasants. Some of the unfavourable characteristics of monsoon are: (i) the breaking of the monsoon rain may be delayed and also exhausted much earlier than usual causing damage to Kharif crops and making the sowing of Rabi crops difficult or uncertain; (ii) there may be prolonged spells of rainy days in the months of June, July and August, when summer crops need plenty of water; (iii) over 75 per cent of Assam's annual rainfall is recorded in the summer season and continuous drought prevails throughout the winter months; (iv) monsoon rain is erratic, sometimes falling in heavy downpours for several days together without any break leading to flood havoc and on the undulating surface causing excessive soil leeching and soil erosion, thereby removing the soil fertility of the land; sometimes the rainfall is very light and discontinuous causing the standing crops to suffer from inadequate supply of water.

In view of the uncertain characteristics of monsoon rains in Assam, there exists a prima facie case for favour of irrigation, even in the summer months, for assured supply of water to the crops. Flood control and protection of surface soil from soil erosion are the two other prime requisites for successful agricultural operations in the State.

Climatic Regions

Mainly on the basis of rainfall, the State may be divided into six climatic zones: (1) Upper Brahmaputra Valley, (2) Central Brahmaputra Valley, (3) North Bank Plains, (4) Lower Brahmaputra Valley, (5) Barak Valley, and (6) Hills Region.

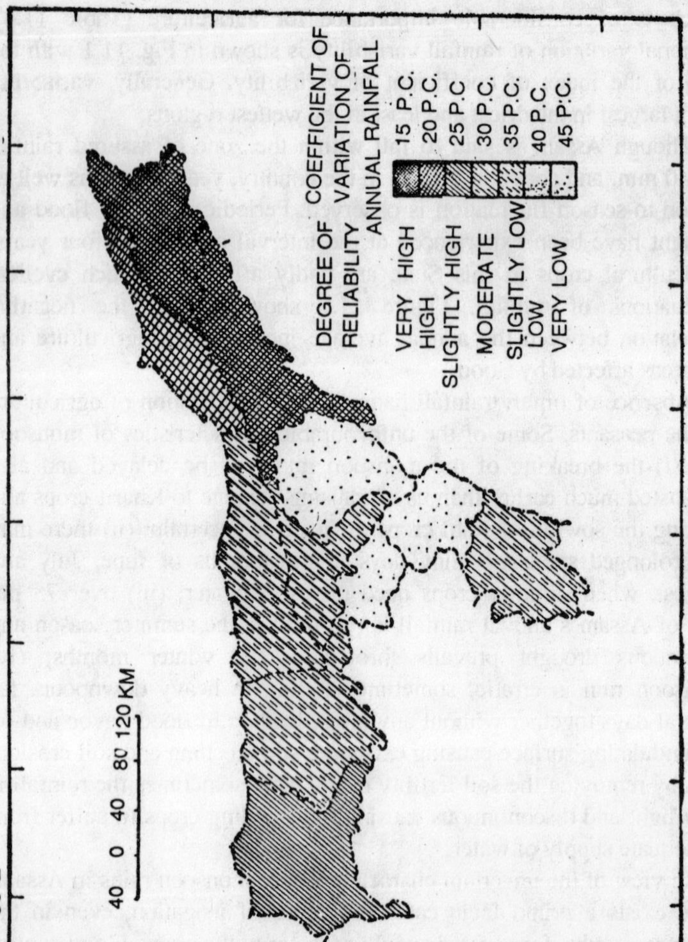

FIG. 11.1: Assam rainfall reliability map.

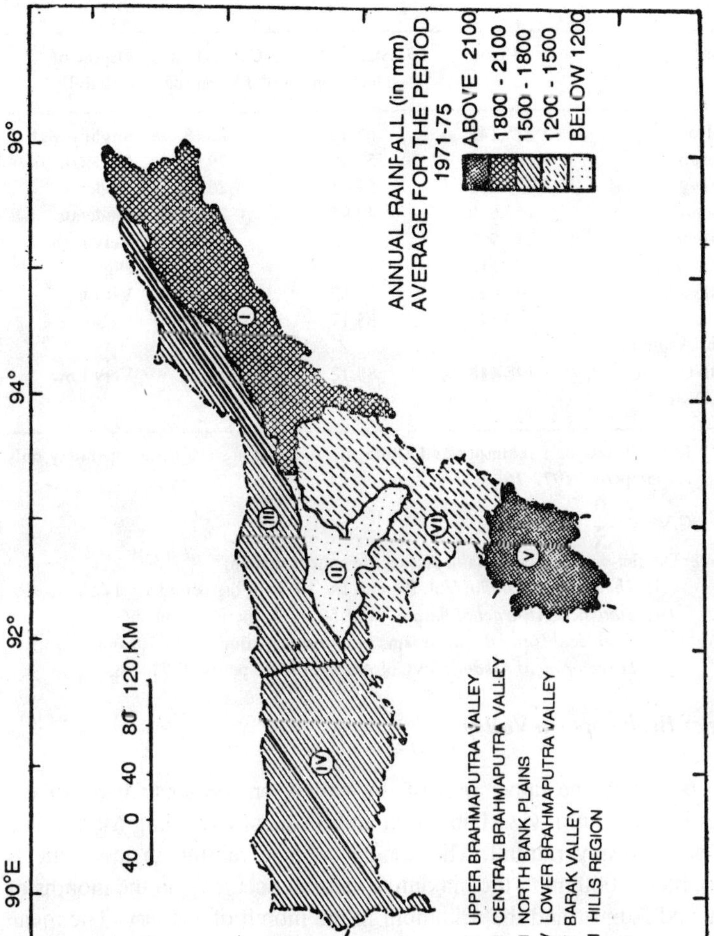

FIG. 11.2: Assam's climatic regions.

TABLE 11.5: **Index of degree of rainfall reliability in Assam during 1951–76**

(rainfall in cm.)

District	Mean	Standard Deviation	Coefficient of Variation †	Degree of Reliability
Goalpara	275.142	65.17	23.68	Slightly high
Kamrup	181.242	53.48	29.51	Moderate
Darrang	223.270	59.04	26.44	Moderate
Nowgong	147.650	43.95	29.76	Moderate
Sibsagar	212.999	29.36	13.78	Very high
Lakhimpur*	251.512	48.35	19.22	High
Dibrugarh*	202.024	20.82	10.31	Very high
Cachar	333.57	85.19	25.54	Moderate
Karbi-Anglong North Cachar Hill District	198.448	80.12	40.37	Very Low

* Rainfall data of Lakhimpur and Dibrugarh districts are available separately only for the period 1972-76.

† $\text{C.V.} = \frac{\delta}{\bar{X}} \times 100$

Source: District-wise rainfall data of Assam are collected from:

(i) *The Census of India*, Vol. III, Part IA, 1961 for the period 1951–59.
(ii) *Statistical Abstract of Assam*, 1967-68 for the period 1960–66.
(iii) *Statistical Hand Book*, Assam, 1973 for the period 1966–70, and
(iv) *Directorate of Assam*, Govt. of Assam, for the period 1971–76.

Upper Brahmaputra Valley

The region in the upper part of the valley, on the south bank of the river Brahmaputra, viz., Dibrugarh and Sibsagar excluding Majuli, is a region of heavy rainfall. The annual average rainfall in this zone is more than 2,000 mm., the maximum rainfall occurring in the months of July and August and the minimum in the month of January. The mean maximum temperature rises to 32°C in July and August at Sibsagar and 30°C at Dibrugarh. The mean minimum falls to 10°C at both stations. The annual mean per cent of relative humidity stands above 84 throughout the whole region. The onset of monsoon is earliest in this region like that in the Barak Valley and the number of rainy days in the winter is highest here. The damp and comparatively cool climate of this zone is ideal for tea, rice and sugarcane cultivation. The zone is relatively immune to devastating floods and drought except river erosion at Dibrugarh and Neemati near Jorhat. Figure 11.1 reveals clearly that rainfall reliability is very high in this climatic zone.

Central Brahmaputra Valley

This zone consists of Nowgong district encircled by the hills of the Meghalaya Plateau in the south and the west, and the Rengma Plateau in the east. On the north there is only the river Brahmaputra flowing. Because of this physiographic peculiarity, the average annual rainfall in this district is lowest among the districts of Assam, i.e. only 1,206.37 mm. during the period 1965-76. The highest rainfall occurs in the month of June and the minimum in January. The temperature rises to a mean maximum of 32°C in the months of July, August and September and falls to a mean minimum of 11°C in January. From Gauhati eastward along the foothills of the Meghalaya Plateau, rainfall decreases reaching its minimum in Lanka — Lumding region of South Nowgong. This region lies in the rain-shadow zone behind the Meghalaya Plateau which stands directly in the path of the south-west monsoon from the Bay of Bengal. The Nambar forest in the western part of the Sibsagar district also experiences this rain-shadow effect. In order to counteract the deficiency of climogenetic water supply, irrigation is inevitable for successful agriculture in these rain-shadow areas.

The northern part of Nowgong is usually inundated by floods during the rainy season. The low-lying plains produce a good variety of jute. Rice and jute are the two important crops grown in large scale in this climatic zone. This climatic region falls in the moderately reliable rainfall zone (Fig. 11.1).

North Bank Plains

This is an elongated zone extending from the Barandi on the west to the river Dehang river on the east in the north of the Brahmaputra. The zone includes Darrang district, Lakhimpur district and Majuli, the biggest river island inside the Brahmaputra. Here the average annual rainfall is more than 2,000 mm., but it becomes gradually heavier as one goes from west to east. The maximum rainfall occurs in the months of July, August and September and the minimum in January in the eastern part of this zone, while the western part receives the highest amount of rainfall in June, July and August and almost no rainfall in December. Temperature rises to a mean maximum of 32°C in July and August and falls to a mean minimum of 11°C in January.

The foothills of the Himalayas are covered by deep forests.

Towards the south of this belt there are several tea estates extending from Brandi to the river Subansiri. The central belt composed of alluvial soil is rich for rice cultivation and human settlement. The low-lying areas near the river Brahmaputra, the areas between the Subansiri and the Brahmaputra, and in Majuli island are inundated by floods almost every year. In the low-lying area of the western part, jute is cultivated and in the lowlands of the east "bao" rice is important. A comparison with Fig. 11.1 reveals that rainfall is slightly reliable in Lakhimpur and less reliable in Darrang district.

Lower Brahmaputra Valley

This region consists of Kamrup and Goalpara districts. The average annual rainfall is 2,800 mm. in Goalpara and 1,800 mm. in Kamrup. It is lowest in the south-eastern Kamrup around Gauhati with an annual average of only 1,600 mm. Incidence of rainfall increases towards north and west. The annual average rainfall at Dhubri is 2,500 mm. December and January are the driest months. Rainfall is heaviest in the months of June, July and August. From March to October the mean maximum temperature remains above 30°C at Gauhati and falls to a mean minimum of 8°C only in the month of January. But Dhubri receives the minimum temperature of 11°C in January, while the maximum above 30°C remains from March to October. Rainfall is highly reliable in Goalpara, but in Kamrup it is less reliable (Fig. 11.1).

The northern foothills are covered with dense forests and tall grasses. Intensive cultivation is done by the peasants in the central portions of the plain towards the north and the south of the Brahmaputra. In the low lying areas near the banks of the Brahmaputra, jute is cultivated. In the winter dry season, wheat is also grown. Though this climatic region is agriculturally rich, the crops in a major portion of the region are devastated annually by the floods created by the tributaries with shallow beds coming down from the high Himalayas with huge volumes of water and a load of sediments.

Barak Valley

The region covers only Cachar district, completely cut off from the Brahmaputra valley by the Meghalaya Plateau and the Barail Range of North Cachar Hills on its northern side. This small plain is the upper part of the great Surama-Meghna river basin and is bounded on the east by the Manipur hills and on the south by the Mizo Hills. This is the region of heaviest rainfall in Assam, the average annual rainfall being more than 3,000 mm. The rainfall is not well distributed throughout the year. June, July and August are the months of heavy and continuous downpours, while the winter season is almost dry. Reliability of rainfall is not high unlike the Upper Brahmaputra Valley, another region of heavy rainfall (Fig. 11.1). Frequent floods occur in the rainy season in the central alluvial plain, as a result of heavy rainfall and shallow river beds of both Barak and Kushiara. A mean maximum temperature of 30°C is recorded in the month of August and the mean minimum of 11°C in January.

The hill slopes are dotted with tea estates and the central alluvial plain is dominated by the mono-crop of rice cultivation.

Hills Region

The region comprises the two hill districts, viz. Karbi Anglong and North Cachar (N.C.) Hills. Being situated in the rain-shadow areas, the region with low hills and peneplain areas gets lesser amount of rainwater in comparison to other regions of the State except the central Brahmaputra valley. The annual average rainfall is 1,500 mm. the maximum fall occurring in the months of July and August. The winter is almost devoid of rain. Reliability of rainfall is lowest in this region (Fig. 11.1). Due to relief, temperature is mild in this region in comparison with the plain areas. Rice and cotton are the two important crops grown by the hill tribes by the method of jhuming.

Conclusion

From the study of the climatic pattern of Assam, it becomes evident that there are wide diversities of the distribution of various elements of weather and climate in respect of both time and space within the State and, therefore, the State may be regionalised into some climatic regions on the basis of climatic homogeneity and variability. It is interesting to find that these climatic regions almost closely coincide with the cropping patterns of different regions. These climatic regions may, therefore, be considered as agro-climatic regions.

Besides, the study reveals the flood-affected areas, flood-immune area and also the areas in urgent need of irrigation facilities. Moreover, it becomes apparent that the existing agricultural pattern in the State is based on greater reliance on monsoonic climatic condition, which is usually erratic and uncertain. Since natural climatic conditions cannot be controlled easily, measures must be initiated to counteract the ill affects of climate by providing irrigation water to the agricultural fields, flood control and protection of soil erosion and so on.

REFERENCES

Borthakur, M., 1968 : "Some Aspects of Weather in the Brahmaputra Valley", *Jr. Assam Science Society*, 2.

Das, H. P., 1970 : *Geography of Assam*, National Book Trust, New Delhi.

Directorate of Economics and Statistics, Govt. of Assam, 1970-71 : *World Agricultural Census*, Assam.

Goswami, P. C., 1963 : *Economic Development of Assam*, Asia Publishing House, New York.

Schimper, A. F., 1903 : *Plant Geography upon a Physiological Basin*, Oxford, pp. 35-51.

Trowartha, G. T., 1954 : *An Introduction to Climate*, 3rd ed., McGraw Hills Book Company, INC, New York, pp. 382-83.

12

FROST HAZARD IN NORTH-EAST CENTRAL ALBERTA

R. HAYTER

TEN YEARS ago Longley (1967) drew attention to the problems of defining the frost-free period in Alberta. In the present paper we discuss further problems that arose in a study carried out in the late 1960s of the incidence of frost and its effects on agriculture in the continuously settled part of the province north-east of Edmonton. The study area is characterised by undulating terrain which is deeply incised in restricted areas by the rivers Athabasca and North Saskatchewan, and by their major tributaries. Numerous lakes and extensive muskeg areas occur, especially in the north of the area which is drained by the ill-defined headwaters of the Churchill River system.

Throughout the study area, as elsewhere along the northern frontiers of the agricultural settlement in western Canada, frost is both a constant limitation to the range of agricultural practices and a killing hazard which can occur sporadically during the growing season. The untimely occurrence of frost results directly in substantial losses of crops in occasional years, as in 1968 when the sample of farmers interviewed in the study area estimated that 10 per cent of the area sown with cereals had been destroyed, and grain yield reduced overall by 50 per cent as a result of widespread frost in early August. Additionally, frost reduces crop quality in many years.

The chance of frost damage can be considered in terms of the probability of frost occurrence on particular dates in spring and fall, as determined from long-run temperature records of meteorological stations at point locations. Such meteorological probabilities of frost occurrence refer strictly to specific points, and their extrapolation to wider areas depends on the representativeness of individual meteorological stations. In fields of growing crops the chance of damage depends not only on microvariations in minimum temperatures, but also on differences in plant susceptibility, reflecting stage and patterns of growth which, in turn depend on the previous adequacy of heat and moisture supplies, soil fertility and cultural practices.

In a broader sense frost can be considered a hazard to the farmer, for it is a cause of crop losses, and agricultural opportunity is restricted by the short frost-free season. However, the spatial extent of frost damage cannot readily by demonstrated, for severity of damage varies, and the damage itself is often not easily isolated from other causes and is sometimes incidental to other loss-inducing causes, such as drought or low levels of plant nutrition. The response of individual farmers in the study area to the frost hazard is a function not only of their awareness and knowledge of the incidence of frost and its effects, but also of their assessment of other physical factors and the broad range of economic factors which govern profitability. Relations between landuse and frost occurrence are, therefore, complex, but their understanding requires an adequate knowledge of the spatial and temporal incidence of frost in the area. Hence, an attempt has been made to study it in Central Alberta (Fig. 12.1)

The basic agro-climatological elements of the northern areas of Alberta are well-known, and their broad outlines distinguished spatially (for example, Longley, 1968). The limited growing season and the susceptibility to summer frost is broadly a function of a northerly continental situation. The absence of a mountain barrier to the north allows easier influx of cold arctic air, while distance from warm areas reduces heating by advection. In contrast, advection frosts occur when cold-air masses move into the region from the north. Such frosts are frequently widespread and can be associated with strong winds, as was the case on June 10, 11 and 12 of 1969 when cold air from a high-pressure system located to the west of the Hudson Bay reduced temperatures by as much as 12°F below freezing, and even caused frost in the southern part of the Province. Radiation frosts occur normally under stable high-pressure conditions, when nocturnal radiation reduces the temperature below freezing. These frosts are associated with clear skies, calm winds and, frequently, an inversion of temperature. Spatially they may be restricted, occurring especially in lower areas into which cold air can drain. Frosts can also result from a combination of these two types when, for example, cold air blows in during the day and the wind drops at night, and radiation effects emphasise the existing low temperatures.

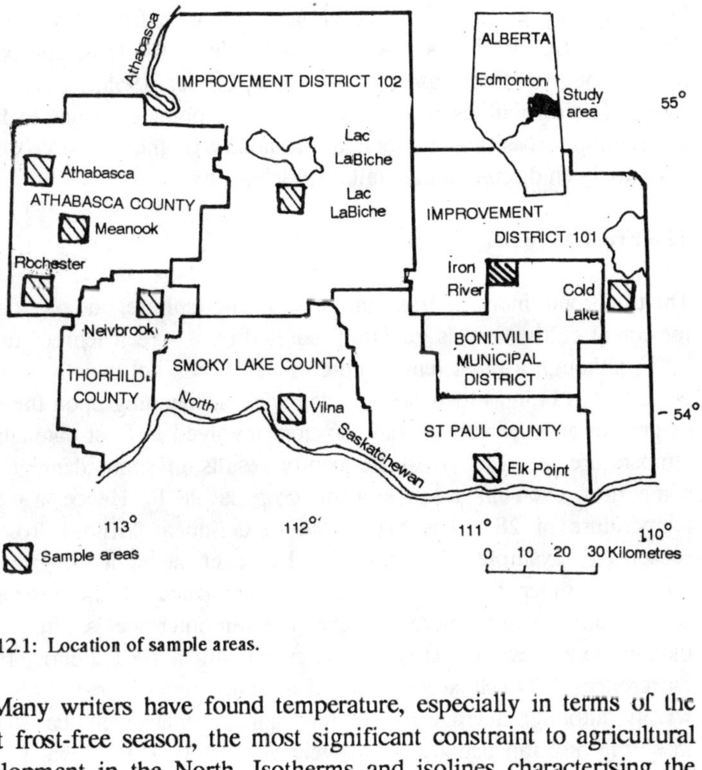

FIG. 12.1: Location of sample areas.

Many writers have found temperature, especially in terms of the short frost-free season, the most significant constraint to agricultural development in the North. Isotherms and isolines characterising the length and warmth of the growing season, and the shorter frost-free season have been used to delimit both the potential and contemporary cultivation limits of grain, especially wheat (Proudfoot, 1974). Williams (1969) argued that wheat should mature as far north as Wrigley at 63° 12′ N, along the Mackenzie, the same northern terminus as that suggested by Unstead in 1912. However, the entire area to the north of Edmonton has been characterised by Chapman and Brown (1966) as facing a serious frost hazard, since north-eastern fringe of cultivation in Alberta is receiving less heat in terms of accumulated temperature than any other agricultural area in Canada. In contrast, Carder (1965) and Guitard et al (1965) virtually dismissed the problems of insufficient heat and the risk of frost, and stressed the significance of annual variations in moisture in causing yield variations.

Agro-meteorological interpretations of the northern fringe areas are not, therefore, consistent. Those who favour the potential of the North for agriculture either ignore the variable factor of frost (for example, Unstead, 1912; Williams, 1969), or use 28°F as the most appropriate limit of the frost-free season (for example, Carder, 1965; Guitard et al, 1965). The difficulties of extrapolating from point locations to larger surrounding areas, and the annual variability of the frost-free season have not been discussed in detail.

The Frost-Free Season

The traditional index of frost limitation for agriculture, and particularly for cereal cultivation, is generally defined by a screen temperature of 32°F, although a plant temperature of 32°F does not usually result in cereal crops being killed or even damaged. Depending upon the stage of growth and the various other factors involved in frost damage, the temperature at which frost frequently results in some damage, and sometimes even in total loss of crop is 28°F. Hence a screen temperature of 28°F has been used to define a "killing" frost-free season (for example, Carder, 1965). However, at night the ground is normally colder than the temperatures indicated in the instrument shelter, and the temperature at the plant-air interface is often lower than in the air several feet above the crop (Geiger, 1961; Pelton, 1967). Differences between screen and grass minimum temperatures vary widely, although average figures for Canadian meteorological stations in spring and fall usually lie between 5° and 10°F (Canada, D.O.T., 1968). At Beaverlodge, in the Peace River region, which was one of the few stations north of Edmonton reporting grass minimum temperatures at the time of our study, grass minima were on average 8°F lower than screen minima during the period May to October, 1968. The data do not show a perfectly normal distribution, although they are strongly clustered around the mean for each of the months, and for the period as a whole (Figs. 12.2 and 12.3). During July and August, 16 frosts were recorded at ground level, but none at screen level, despite the widespread occurrences of frost damage in early August right across the northern part of the Province. There is thus good reason for accepting the commonly used definition of frost as a screen temperature of 32°F.

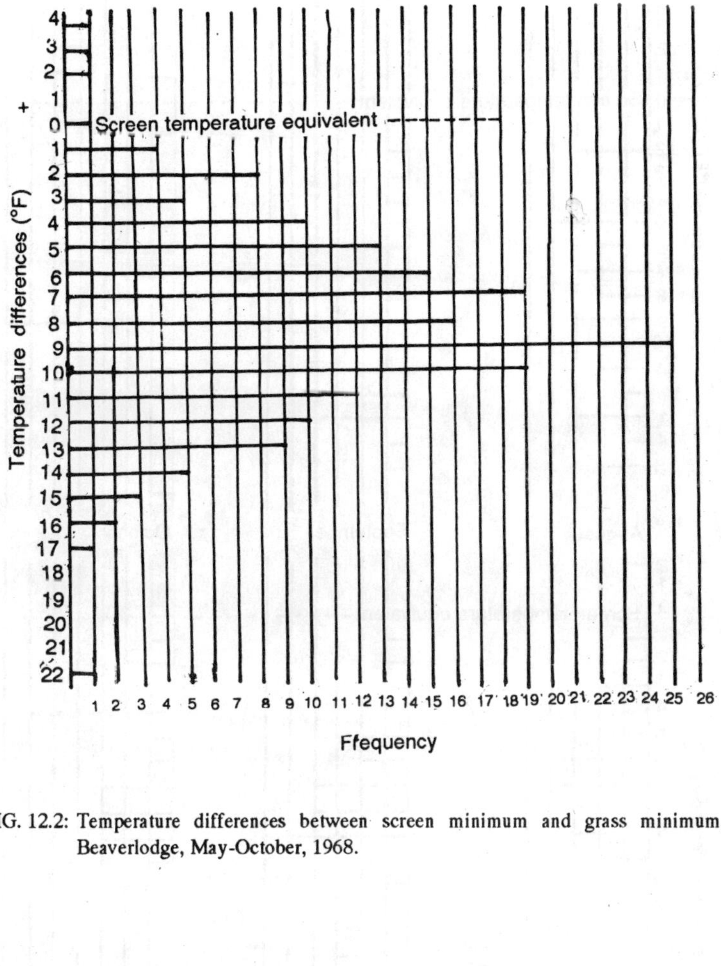

FIG. 12.2: Temperature differences between screen minimum and grass minimum, Beaverlodge, May-October, 1968.

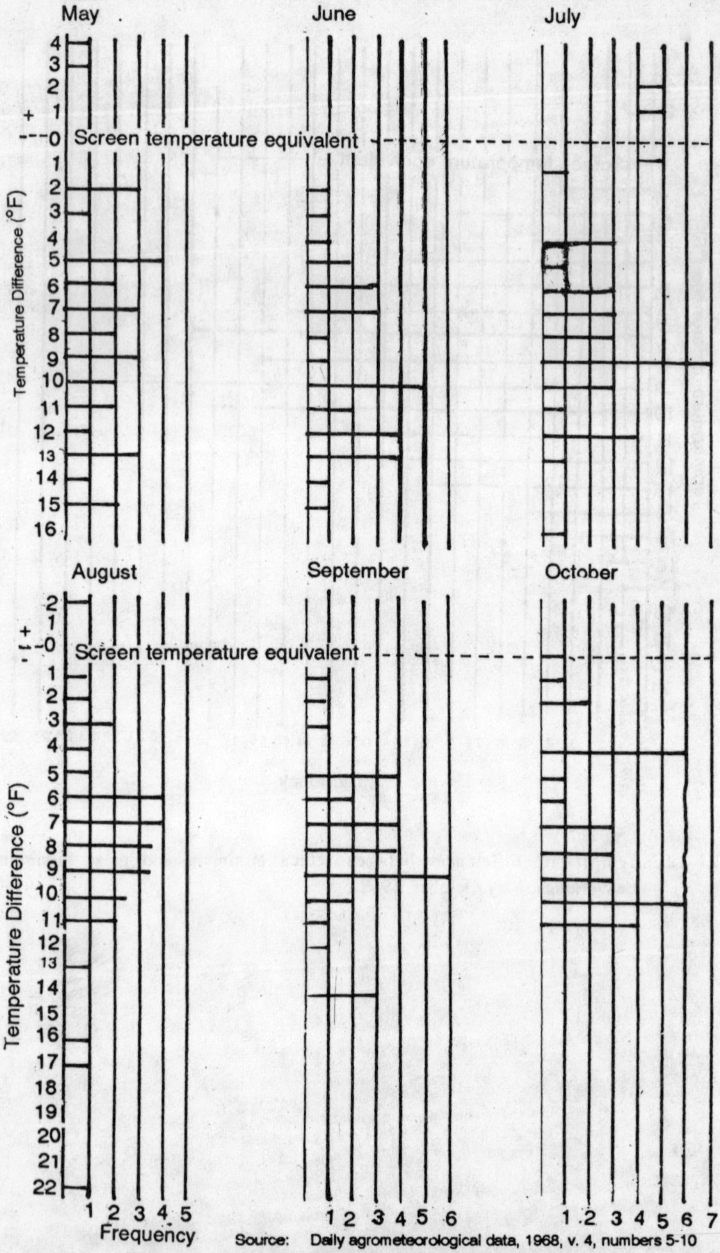

Source: Daily agrometeorological data, 1968, v. 4, numbers 5-10
Dept. of Transport, Meteorological Branch

FIG. 12.3: Temperature differences between screen minimum and grass minimum by individual months, Beaverlodge, May-Oct, 1968.

On the basis of this definition Reed (1916), Bennett (1959) and Leonard and Martin (1963) have suggested a mean period of 90–100 frost-free days as the minimum to enable cereals to be grown successfully. Table 12.1 lists the average lengths of the frost-free season and the respective dates of the last spring and first-fall frosts during various periods for the nine meteorological stations in the study area. Median values have also been calculated for the period 1959–1968, but the difference between means and medians is small, being in all cases less than six days, so that the mean can be taken as the 50 per cent level of risk.

TABLE 12.1: Average frost-free periods

	1931—1960	1951—1964	1959—1968	Standard Deviation
Athabasca		84 (June 8-Aug. 31)*	106 (May 27-Sept. 10)	23
Meanook		117 (May 21-Sept. 15)	124 (May 17-Sept. 18)	13
Rochester		79 (June 12-Aug. 30)	73 (June 13-Aug. 25)†	27
Newbrook		81 (June 6-Aug. 26)	67 (June 14-Aug. 20)	23
LacLaBiche		117 (May 22-Sept. 16)*	101 (June 1-Sept. 10)	19
Vilna		85 (June 9-Sept. 2)	75 (June 14-Aug.28)	23
Iron River	77 (June 10-Aug. 27)	103 (May 28-Sept. 8)	105 (May 25-Sept. 7)	11
Cold Lake		116 (May 23-Sept. 16)	117 (May 23-Sept. 17)	20
Elk Point	69 (June 13-Aug. 22)	95 (June 2-Sept. 5)	105 (May 26-Sept. 8)	9

* These stations are not the same as those used for the 1958—1968 period, although in the same locality. In both cases the stations were dismantled in the early 1960s.

† The station at Rochester ceased recording in 1968.

Source: Climatic Normals, Vol. 6, Frost Data, Canada, Deptt. of Transport, Meteorological Branch, 1968, p. 4; Longley, 1967, pp. 244-245; and Canada, Deptt. of Transport, Meteorological Branch, Monthly Records, 1959—1968.

Considering the mean and median values, only two stations, Cold Lake and Meanook, record a frost-free period significantly longer than 100 days during the various periods for which data are available. Three stations, Newbrook, Vilna and Rochester, are well below the required minimum. For the shorter period 1959–1968 Lac La Biche records only 101 days, although another station in the locality records 117 for the longer 1951–1964 period. Elk Point and Iron River both record 105 days during the 1959–1968 period, but Elk Point records only 95 during the 1951–64 period, and both record much lower values during the longer 1931–60 period. All the frost-free periods except those for Cold Lake and Meanook become even more marginal for successful cereal production when it is considered that they constitute average values, and thus there is only an approximately 50 per cent chance of a frost-free season as long or longer. The frost risk facing the farmer is, of course, increased by the impossibility of being able to predict, in any one year, either the length of the frost-free season or the dates of the last spring or first-fall frost. There is considerable year-to-year variation in the length of the frost-free season, in its dates of occurrence, and in the severity of the frosts which limit the season, in spring and fall. During the 1959–1968 period the frost-free season recorded in the study area varied from 23 days at Newbrook to 151 days at Cold Lake. For individual stations, standard deviations about the mean number of frost-free days are large, although the ten-year period for which they are calculated is short. Additionally there is no consistent relationship between the length and dates of occurrence of the "killing" frost-free season and the frost-free season, as shown in Figs. 12.4a, b, ... Longevity or shortness of the frost-free season based upon 32°F does not imply the same for the "killing" frost-free season. The diagrams illustrate clearly the shortness of the frost-free season and its large variation from year to year. During the period 1959–1968 greatest variation in the length of the frost-free season for individual recording stations was 85 days for Athabasca, Rochester and Newbrook, and least variation was 95 days for Elk Point. Clearly the variation and lack of predictability of the occurrence of frost and of the length of the frost-free season can cause a considerable risk to the farmer in the study area. For the two stations of Iron River and Elk Point with long records the dates of last spring frost and first-fall frost have been calculated for given risks as shown in Table 12.2. The value of such risk tabulations is reduced, however, because of temporal climatic variations (even if climatic cycles as such cannot be recognised) and because of changes in settlement and vegetation around meteorological stations. An indication of the temporal variation in the lengths of the frost-free season in this area is given by comparing the 1931–1960 means of 77 and 69 days for Iron River and Elk Point with those for 1951–1964 of 103 and 95 days, and for 1959–1968 of 105 days for both stations. The study period 1959–1968 includes part of the 1951–1964 period when warming trends were evident in Alberta (Longley, 1967). Moreover, problems in calculating frost probabilities arise not only because of temporal variations, but because of

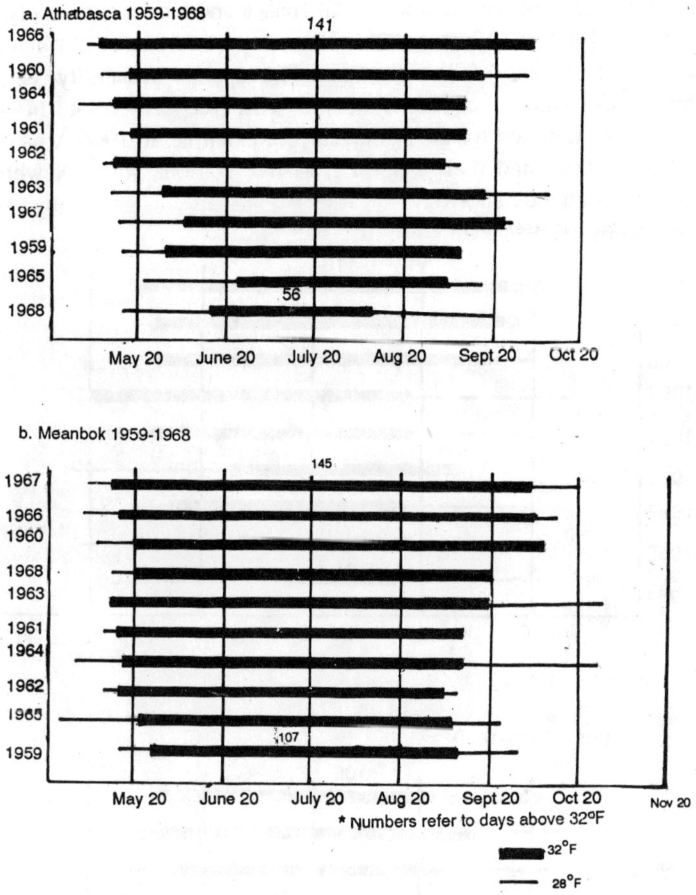

FIG. 12.4: Frost-free season.

difficulties in extrapolating spatially from individual recording stations. Both at a local and regional level there can be substantial variations over relatively short distances; for example, the Meanook and Rochester stations were situated on the same undulating plateau surface only 19 miles apart, yet the average frost-free season differed during 1959–1968 by 51 days. Similarly, the differences in the average length of the frost-free seasons at Athabasca and lac La Biche between the 1951–1964 and 1959–1968 periods are probably to be attributed in part to relocation of the stations. The spatial variability is also emphasised when the length of the frost-free season at different stations is examined for the same year; for example, in 1966 Athabasca and Iron River had their longest frost-free seasons while Newbrook and Vilna had their shortest.

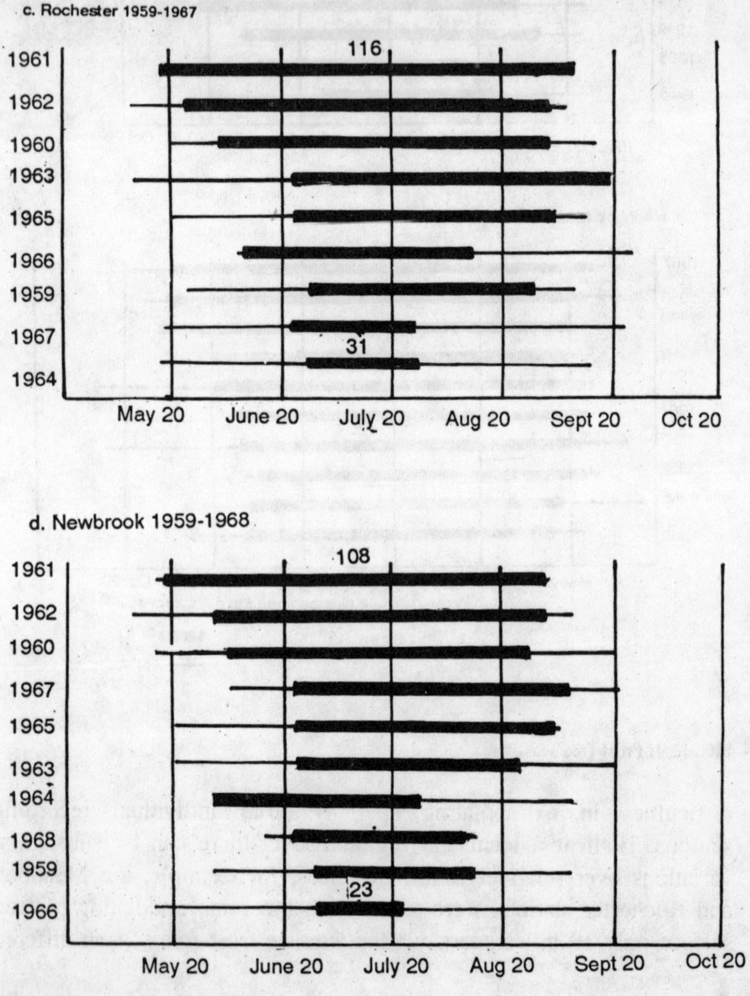

FIG. 12.4: *(Contd.)*.

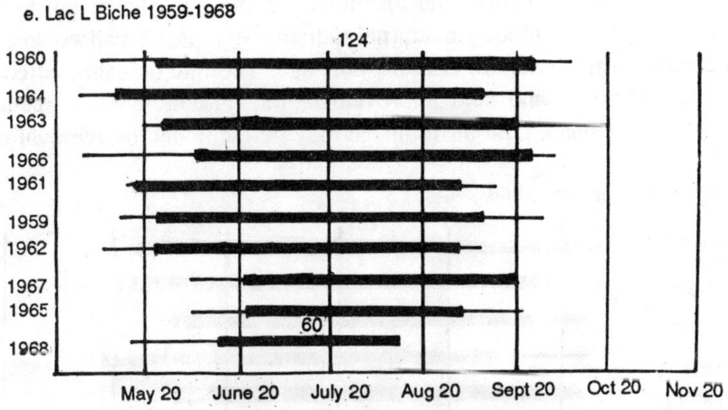

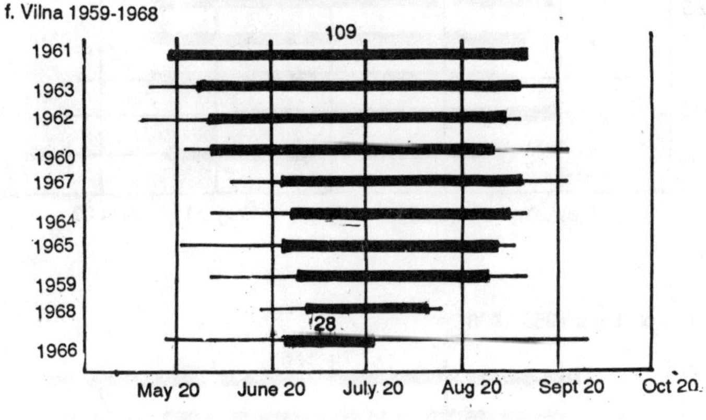

FIG. 12.4: *(Contd.)*.

Micro-climatic Variations and Station Representativeness

In the study area, with its varied topography, the principal microclimatic variations are associated with cold-air drainage resulting in temperature inversions, the presence of large water bodies which may completely counteract the mechanism of cold-air drainage, and vegetation, the effect of which varies according to type. Muskeg, which is associated with cold, damp peaty soils, results in, and in turn helps to maintain, frost-prone localities. On the other hand, tall trees and garden shrubs modify nocturnal radiation in highly localised areas, and may form barriers to cold-air drainage. The microclimatic effects of tree clearance and land improvement for agriculture are variable. On the other hand, cold-air drainage may be facilitated by removal of

g. Iron River 1959-1968

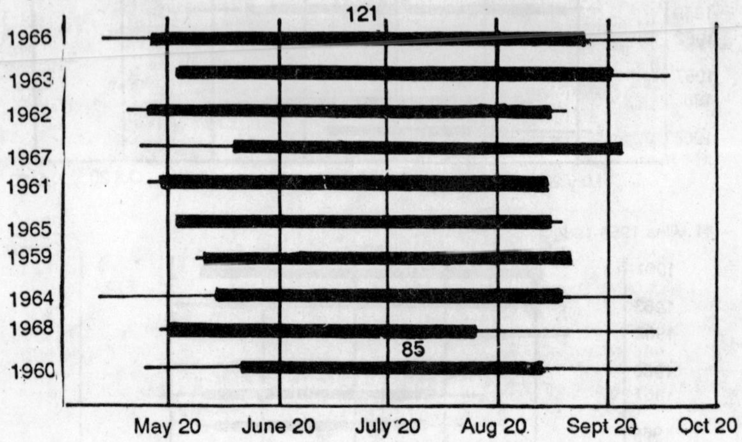

h. Colk Lake 1959-1968

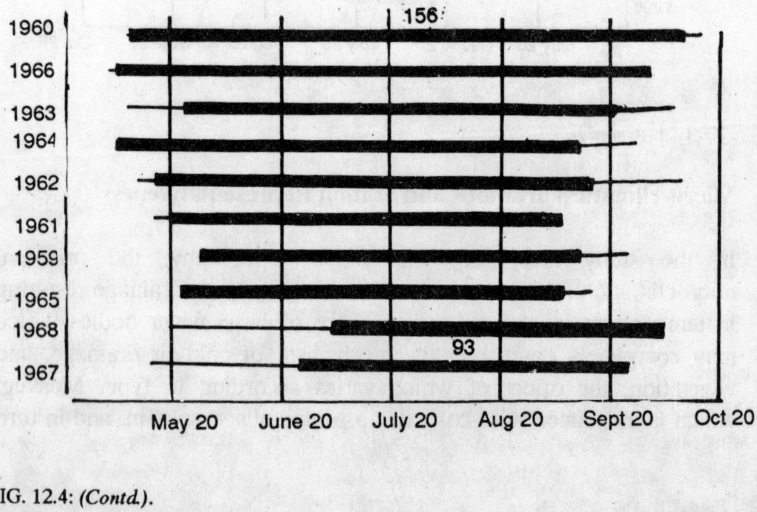

FIG. 12.4: *(Contd.)*.

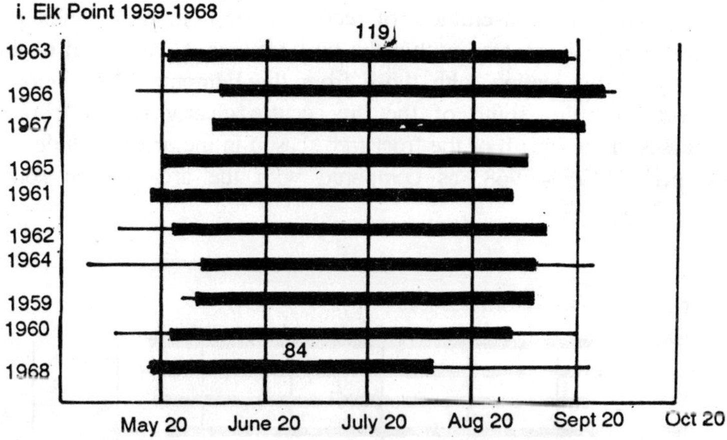

FIG. 12.4: *(Contd.).*

TABLE 12.2: Spring date of last frost and fall date of first frost for a given risk

	Spring					
	1931—1960			1959—1968		
	10%	50%	90%	10%	50%	90%
Iron River	June 25	June 6	May18	June 7	May 25	May 14
Elk Point	June 29	June 9	May21	June 5	May 26	May 18
			Fall			
Iron River	Aug. 7	Aug. 31	Sep. 25	Aug.31	Sep. 7	Sep. 22
Elk Point	Aug. 6	Aug. 29	Sep. 21	Aug.30	Sep. 8	Sep. 29

Source: *Climatic Normals*, Vol. 6, Canada, Deptt. of Transport, Meteorological Branch, 1968; and Canada, Deptt. of Transport, Meteorological Branch, Monthly Records, 1959—1968.

trees. On the other, the clearance and land drainage of muskeg tends to lengthen the frost-free season, as the organic matter content of the soil decreases (compare Currie, 1959; Taylor, 1967). Tall trees and shrubs, particularly in association with a well-cultivated garden, protect thermometers placed near them against cold air drainage, and also form a barrier to radiation heat losses. The temperature recordings of Elk Point, Athabasca and Iron River are all influenced in this way. Long-term records from stations in such situations are also difficult to evaluate, because growth of shelter belts over time may alter the micro-climates significantly and thereby mask any more general temporal variations. Within the study area the "heat-island' effect of urban places is not important, although the growth of Edmonton has clearly reduced the usefulness of records from Edmonton Municipal Airport which is situated within the built-up area. Comparison of the records of this station with those from the Edmonton International Airport 15 miles south of the city centre clearly shows marked increases in the length of the frost-free season in the urban area during the period 1959–1968, as compared with the area to the south (Fig. 12.5).

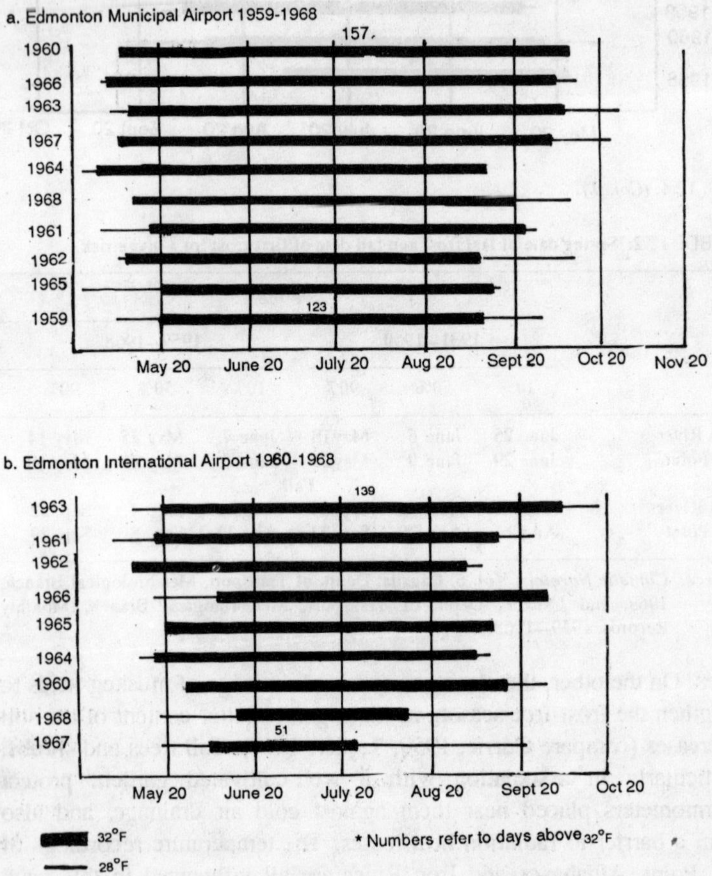

FIG. 12.5: Frost-free season.

TABLE 12.3: Results of temperature traverses

	Station Site	Temp. of Station	Highest Temp. Recorded	Lowest Temp. Recorded	Temp. Variation	Comments	Conclusion
1	2	3	4	5	6	7	8
Athabasca	On a slight rise with large windbreak (line of trees) to north	26.2°F	27.0°F	23.8°F	3.2°F	Variation in minimum temperature probably considerably reduced during course of traverse by low-lying mist.	Some indication that the frost records of this station tend to be too favourable for the sample area.
Meanook	On the top of a prominent knoll, one of the highest points in sample area	27.5°F	27.5°F	19.4°F	8.1°F	Along the traverse route, temperatures generally at least 4°F lower than station temperature and some inversions 8°F lower.	Because of site factors, particularly cold air drainage, this station should be ignored in a regional assessment of the frost hazard.
Rochester	Near the foot of a gradual slope in an undulating plateau surface	36.3°F	37.6°F	30.7°F	6.9°F	Temperature variations were reduced by wind gusts of 10 m.p.h. Even so a couple of marked inversions were recorded as a result of cold-air drainage.	Representative of much of sample area except for the higher plateau levels.

(*Contd*)

(*Contd*)

1	2	3	4	5	6	7	8
Newbrook	On level land with no visible slope changes	23.0°F	35.0°F	18.5°F	17.5°F	Over most of area, temperatures were similar or lower than those recorded by station except for a ridge of high land on the north-eastern boundary of sample area.	Micro-climatic factors of station site and results from traverses suggest this station is a reliable index of frost hazard for most of area. Mechanism of cold-air drainage may operate on an extensive scale in this area.
Lac La Biche	On top of a knoll next to a lake (Lac) La Biche) with a windbreak to the north.	23.0°F	24.8°F	16.2°F	8.6°F	At only two places along the route were temperatures similar or higher than the station.Elsewhere temperatures were generally 4°F lower than station.	Micro-climate factors of station site have resulted in this station recording a too favourable impression of frost hazard.
Vilna	Situated on a slight slope	27.1°F	27.1°F	18.5°F	8.6°F	Temperatures similar to stations along most of route except for a couple of marked inversions in shallow hollows.	Records must be considered representative of most of sample area.

(*Contd*)

(Contd)

1	2	3	4	5	6	7	8
Iron River	On a small rise in a well-cultivated garden in close proximity to tall trees and shrubs.	23.0°F	23.9°F	14.0°F	9.9°F	Temperatures mostly 2°F lower than station temperature in certain shallow depressions by as much 10°F.	Micro-climatic factors of station site result in underestimation of frost hazard for study area.
Cold Lake	On flat land (air base location) near to large lake.	24.2°F	22.7°F	13.8°F	8.9°F	Only alongside shores of Cold Lake was temperature close to that of station. Elsewhere temperatures generally 4°F or more below station temperature.	Like Meanook, this station is representative of only highly localised areas and should be ignored in a regional assessment of frost hazard.
Elk Point	In a well-cultivated garden.	23.9°F	25.7°F	20.7°F	4.9°F	Results of traverse made invalid by incoming warm air mass.	Probably representative of much of sample area.

Because of these various micro-climatic factors it is necessary to determine the particular localities for which each station is representative, and evaluate which stations are most representative of the study area as a whole. In order to assess the micro-climatic effect upon minimum temperature distribution, and the representativeness of each station, a series of temperature traverses was conducted in the fall of 1969. These used the same methods, instruments and procedures as had been used in the topographically similar Springdale area, west of Edmonton, by Longley and Louis-Byne (1967). Traverses were conducted on calm, clear nights between one hour before and half-an-hour after sunrise, when temperatures are generally both at a minimum and relatively constant. As in the Springdale area, the consistency of the temperature patterns on clear nights enabled the general patterns of temperature to be established for the routes, and hence the variations between the stations and surrounding areas to be determined. From the results of the traverses in each sample area, maps and cross-profiles relating temperature distribution to topography, and to the temperature recorded by each station were drawn. Details have been published elsewhere (Hayter, 1972) but the major results are given in Table 12.3. Microvariations are clearly more significant than any broad regional ones within the study area. However, it is difficult to translate the minimum temperature variations recorded on the traverses into lengths of frost-free seasons, although Longley (1967) noted that a change of 1°C in the mean minimum causes a significant difference in the frost-free season. An indication of the extent of this difference is illustrated by the mean minimum temperature records of the Newbrook and Meanook stations (Table 12.4). The mean difference for the five months combined is 5°F, and this is associated with a difference of 57 days in the length of the frost-free period. A difference of only 1-2°F may thus be expected to change the frost-free period by one to two weeks. From a detailed examination of the records of the meteorological stations and from the results of the traverses, a climatological model of the frost hazard for the agricultural area north-east of Edmonton has been derived (Fig. 12.6). Average frost-free seasons of 110 days or more are highly localised: they are limited to the very highest relief levels, as well as to areas along the shores of large lakes, such as Cold Lake and Lac La Biche, and to valleys to large rivers, such as the Athabasca and North Saskatchewan. A period of 90–110 days is also limited to relatively high land from which cold air can drain. For the greater part of the study area a frost-free season of 90 days cannot be relied upon. In certain localised depressions, in the vicinity of muskeg, and in those large steep-sided valleys presently occupied by small streams, frost should be expected in every month of the year. If a single figure for the length of the frost-free season in the study area were to be given it must, to be representative of much of the area, lie within the range of 75-85 days.

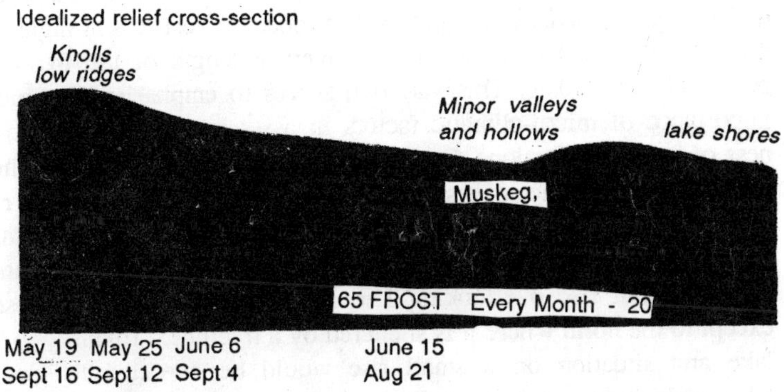

FIG. 12.6: Climatological model of frost incidence in north-east Central Alberta.

TABLE 12.4: Mean minimum temperatures, °F, 1959–1968

	May	June	July	Aug.	Sept.
Meanook	39.4	45.8	52.3	50.3	42.1
Newbrook	35.2	41.5	46.7	44.7	36.5

Source: Canada, Dept. of Transport, Meteorological Branch, Monthly, Records, 1959–1968.

The meanook records can be taken to represent restricted areas with at least 120 frost-free days. The station, situated on top of a prominent knoll that rises steeply above the surrounding plateau surface, gives a very favourable impression of the frost-free season. During the 1959–1968 period the frost-free season was never less than 106 days (1959), and frost did not occur in June, July or August. Frost-free seasons of over 140 days were registered in 1960, 1966 and 1967, and the first fall frost did not occur until early October in three years of the 10 under study. On the basis of the 1959–1968 period there is no risk of frost in spring after May 25, or before September 2 in the fall. On May 15 the risks of spring frosts is 50 per cent and this decreases to less than 10 per cent after May 20. In the fall the probability of frost occurring by September 10 is 50 per cent but 90 per cent probability is not reached until the first week in October. As regards "killing" frosts there is only a 10 per cent risk in spring after May 15, and a 20 per cent risk before September 10 in the fall.

The Lac La Biche station is taken to represent the area with 90-110 frost-free days. During the study period very short seasons were recorded in 1968 (60 days), 1965 (72 days) and 1967 (90 days). Until 1964 the instrument shelter was heavily protected by trees and shrubs which seem to have been significant in reducing heat loss at night, for there is a marked reduction in the average length of the frost-free period after this date. This variation serves to emphasise the crucial importance of micro-climatic factors in assessing the representativeness of individual stations. On the basis of the 1959–1968 record there is a 90 per cent chance of a spring frost on May 15 and by June the risk is still 40 per cent. The probability of "killing" frosts for the same dates is 40 per cent and 10 per cent respectively. The meteorological station is on a small rise overlooking Lac La Biche and is well exposed, except to the north where it is sheltered by a tree line. Proximity to the lake and situation on a small rise would be expected to give a favourable record of the frost-free season, yet even at this station there is still the possibility of a "killing" frost in early June. In the fall there is a 10 per cent probability of frost before August 31, and a 70 per cent probability by September 10. However, there is a 10 per cent risk of a "killing" frost in late August and by September 10 there is a 30 per cent risk of such a frost.

The Newbrook station is representative of low-lying-level areas and the records for 1959–1968 show an average frost-free period of only 67 days. Only in 1961 did this station record a frost-free season of more than 100 days. Year with particularly short frost-free seasons were 1966 (23 days), 1959 (45 days) and 1968 (48 days). A "killing" frost-free period of only 60 days was recorded in 1968, while a 32°F frost occurred in every month except July. However, frosts occurred in July and August in six of the 10 years of the study period. The risk of a spring frost at May 15 is 100 per cent and on June 1 is still as high as 60 per cent. Frosts occurred in 1959 on June 28, and in 1966 on June 29. There is a strong probability of a late June frost in the large areas for which Newbrook is representative, and there is even a 20 per cent risk of a "killing" frost after June 1. In the fall the risk of frost is 60 per cent and of "killing" frost 10 per cent before August 31. By September 10 the respective probabilities are 100 per cent and 60 per cent.

A study of the Newbrook, Lac La Biche and Meanook records indicates that July is the month with fewest frosts. However, July frosts have been recorded by the Newbrook, Vilna and Rochester stations, and, surprisingly, by Cold Lake, so that even in July a farmer in north-east Alberta cannot be certain to miss frosts.

Conclusions

Within the agricultural area. north-east of Edmonton there is no general, regional variation in frost occurrence and in the length of the frost-free period. Instead, micro-climatic variations are significant. Meteorological stations within the area are not sited on the whole range of terrains, and individual stations are representative of only limited areas. During the decade 1959–1968 the recorded frost-free season varied from 23 to 151 days, and the station means for the period from 67 to 124. A single figure representative of much of the area would have lain within the range of 75–85 days. Such a period is marginal for cereal cultivation, and the shortness of frost-free season is certainly a major natural contributor to the economic problems of the area.

REFERENCES

Bennett, M. K., 1959 : "The Isoline of 90 Frost-Free Days in Canada", *Economic Geography,* 35, pp. 41-50.

Canada, Government of, Department of Transport 1968: *Climatic Normals*, Vol. 6, Frost Data, Meteorological Office, Toronto, p. 51.

Carder, A. C., 1965 : *Climate of the Upper Peace River*, Canada, Department of Agriculture, Publication 1224, p. 19.

Chapman, L. J. and Brown M. D., 1966 : *The Climates of Canada for Agriculture*, Report Number 3, Canada Land Inventory, Ottawa, Dept. of Forestry and Rural Development, Queen's Printer, p. 24.

Currie, B. W., 1959 : *The Vegetative and Frost-Free Seasons*, Prairie Provinces and North-west Territories, Vol. 4, Physics Dept., University of Saskatchewan, p. 17.

Geiger, R., 1961 : *The Climate Near the Ground*, Friedrich Vieweg and Sohn, Brunswick, p. 610.

Guirard, A. A., R. L. Taylor, J.C. Brinswade, J.A., Gilbey, J.A., Newman and Tsukamoto J.Y., 1965 : *Growth of Spring Cereals in North-western Canada and Alaska*, Dept. of Agriculture Publication 1220, p. 11.

Hayter, R., 1970 : *The Frost Hazard for Agriculture in North-east Alberta*, Thesis for the degree of Master of Arts, University of Alberta, Edmonton, p. 277.

———, 1972 : *Spatial Variations in Minimum Temperature in North-east Central Alberta*, Environment Canada, Atmospheric Environment, p. 39.

Leonard, W. H. and Martin, J. H., 1963 : *Cereal Crops*, New York, The Macmillan Company, p. 824.

Longley, R. W., 1967 : "The Frost-Free Period in Alberta", *Canadian J. Plant Science*, 47, pp. 239-49.

Longley, R. W., 1968 : *Climatic Maps for Alberta*, Edmonton, Department of Geography, The University of Alberta, p. 8.

Longley, R. W., and Louis-byne, M., 1967 : *Frost Hollows in West Central Alberta*, Canada, Dept. of Transport, Meteorological Branch, CIR 4532, TEC 639, p. 15.

Pelton, W. L., 1967 : "Wheat and Climate", in *Proceedings of the Canadian Centennial Wheat Symposium*, Saskatoon, Western Cooperative Fertiliser, pp. 207-24.

Proudfoot, V. B., 1974 : "The Northern Limit of Agriculture in Western Canada", in *Climatic Resources and Economic Activity*, ed. J. A. Taylor, Newton Abbot, David and Charles, pp. 121-33.

Reed, W. G., 1916 : "The Probable Growing Season", *Monthly Weather Review*, Vol. 44, United States, Dept. of Agriculture, Washington, pp. 509-12.

Taylor, J. A., 1967 : "Growing Season as Affected by Land Aspect and Soil Texture", in *Weather and Agriculture*, ed. J.A. Taylor, Oxford, Pergamon Press, pp. 15-36.

Unstead, J. F., 1912 : "The Climatic Limit of Wheat Cultivation with Special Reference to North America", *Geographical J.*, 39, pp. 347-66 and 421-46.

Williams, G. D. V., 1969 : "Applying Estimated Temperature Normals to the Zonation of the Canadian Great Plains for Wheat", *Canadian J. Soil Science*, 49, pp. 263-76.

13

WEATHER AND OPERATIONAL DECISION-MAKING*

W. J. MAUNDER

IN THE PREFACE to *Quarterly Predictions* published by the N.Z. Institute of Economic Research, it is stated that ". . . economic forecasting is a chancy business for in addition to our own imperfect understanding of the way the economy works, there are also likely to be errors from chance factors such as changes in the weather at home and overseas, from inadequacies in our basic information, and from the unknown effects of political events".

In the light of these pertinent comments, the responsibilities of meteorologists to provide the most appropriate information (past, real time and future) are very clear. Indeed, if meteorology is to serve any purpose, a duty rests upon those meteorologists with skills and reputations in the field to alert the decision-maker in national economic planning of the value of weather information, and to assist him to use such information in the best possible way. However, while it is comparatively easy to make assessments of the *general* relationship, between weather factors and production or consumption, or in some cases prices, the more precise relationships necessary for operational decision-making are much more difficult to formulate (Maunder, 1970). Moreover, even with a perfect weather-economic model, major problems still exist in the acceptance by decision-makers—both within and outside meteorological circles—of this new aid.

The acceptance and use of appropriately weighted meteorological information and forecasts of production, etc., resulting from this information, therefore, offers a challenge to the meteorologist in the

* *Reprinted from Quarterly Predictions,* September, 1977.

fields of marketing and education. The following examples using weather information for planning and production are offered as a first step in this process.

Using Weather Information for Planning and Production

To provide a better planning and production service using weather information, it is necessary first to determine the value of the weather and weather information, and this requires identification of those activities directly or indirectly affected by the weather and weather information; second, meaningful weighted weather indexes must be assessed for these activities. For most purposes, traditional climatological information, produced several weeks after the event, is of limited operational use (whatever its educational or historical value may be or its value in "non-operational" decision-making). To contribute significantly to national economic planning, national meteorological services must, therefore, provide decision-makers with "real-time" weather information, appropriately weighted by economic activities and areas, and this must be available on a time-scale which provides sufficient lead-time for decision-making.

To assist in New Zealand's national economic planning, specific applications of weighted weather information are already being made by the New Zealand Meteorological Service, notably in the dairy, meat and wool, and electricity industries. Variations in all three of these weather-sensitive activities are of prime importance to the overall state of the New Zealand economy, and this is demonstrated by the fact that the current annual value of exports from dairy produce, meat and wool are about 60 per cent of the total value of exports from New Zealand. Similarly, with the value of fuel oil imports required for the generation of electric power totalling about $ 15 per person per year, the value of relevant weather information to maximise the use of water and minimise the use of fuel oil is considerable.

Dairy Production and Planning

Weather-based predictions of dairy production one to two months ahead are provided by the New Zealand Meteorological Service on a regular basis to the Farm Production Branch of the New Zealand Dairy Board. This real-time predicted information enables the Board to more accurately assess the market and shipping requirements, which are directly related to anticipated dairy production and, therefore, allows better planning. This work originated in 1971 when analysis indicated that there was a relatively close relationship between the differences in calculated soil moisture between one season and the next (and the months within these seasons), and the corresponding differences in monthly dairy production. As a result, a dairy production prediction programme was started during the 1972-73 season (Maunder, 1974), and in view of its success it has become operational (Maunder, 1977a).

In the programme, the consecutive daily rainfalls from 23 stations in the North Island are assessed to provide an estimate of the available soil moisture. This monitoring of daily rainfalls enables a real-time assessment of the sequence of days in the month or season when typical soils can be considered to be too dry for pasture growth. Such assessments, together with forecasts of probable conditions for the rest of the month, are then weighted by the distribution of dairy cows and compared with similar information for the same period of the previous season. The resulting differences are then used in the model to give a prediction of the following month's dairy production, in terms of a percentage change from the corresponding month in the previous season.

The first real-time use of the prediction model was made in 1972-73 and in view of its success, the model has been updated each season, so that the predictions for the 1976-77 season were based on data for the 26 seasons — 1950-51 to 1975-76. The differences between the predicted and actual North Island production for March and April, 1977, show an error of –7 per cent in the March prediction, and an error of +11 per cent in the April prediction. However, since in the model these predictions were made at the end of the month *preceding* the month to be predicted, and that the actual production is not known by the Dairy Board until late in the month *following* production, the errors are probably within the requirements for decision-making by the Dairy Board. Nevertheless, it is clear that the model could be improved, particularly in the use of smaller time intervals (possibly 10 days), which would enable a more realistic assessment of the number of days of deficit in the periods preceding production.

The value of being able to predict successfully next month's milkfat production (expressed as a percentage of the previous season's production for that month) is considerable, since in monetary terms a 10 per cent change in production from January to April is worth from $

5 to 8 million *per month.* Table 13.1 shows that *predicted* value difference in milkfat production in March, 1973, was $ 11.3 million less than the actual value of production in March, 1972, whereas the *actual* difference between March, 1972, and March, 1973, had a value loss of $ 13.0 million.

TABLE 13.1: The value* of predicted and actual milkfat production differences between March, 1972 and March, 1973

Region**	Predicted Difference+ ($ m)	Actual Difference ($ m)
North Auckland	–0.9	–3.2
South Auckland	–2.8	–4.2
Bay of Plenty	–0.5	–0.5
Taranaki	–3.8	–3.4
Hawkes Bay/Gisborne	–0.2	–0.9
Wellington/Manawatu/Wairarapa	–0.5	–0.3
South Island	–0.5	–0.3
New Zealand	–11.3	–13.0

* Based on the March, 1972, volume of production and a value of $ 2.50/kg, which is the value if milkfat is converted into cheese and exported (prices as of mid-1977).

** Separate predictions are made for each region and New Zealand.

\+ The difference between March, 1973, and March, 1972.

An analysis of the predictions given to the Dairy Board since the prediction programme was started shows that considerable success has been achieved, particularly when there is a marked change from one season to the next. The early availability of such information enables the New Zealand Dairy Board to more accurately assess the expected dairy production for the season. As a result, the Board is in a better position to plan marketing and shipping activities more effectively.

Electric Power and Planning

The value of relevant meteorological information to the decision-making involved in the correct monitoring, prescheduling and, therefore, generation of electric power in New Zealand is also well established. For example, Press statements such as that issued on March 23, 1976, by the New Zealand Electricity Department that

"Allowing for temperature corrections, electricity used last week was 2.9 per cent below the national guidelines", and "oil generation rose last week, fuel costs amounting to $ 1.1 million", are now common, and have an obvious meteorological basis and implication. Indeed, the New Zealand Meteorological Service has for the last few years been supplying to the Electricity Department vital meteorological information on a routine basis (Maunder, 1971). This information comprises forecasts of weather conditions, especially temperatures, and detailed analyses of temperature departures from the average, weighted according to the distribution of the human population, for the immediate past.

The special weather forecasts for the Electricity Department are issued every morning, and comprise : (1) general forecasts for each Island, (2) temperature forecasts for noon, and 5 p.m. today, and 8 a.m., noon and 5 p.m. tomorrow for Auckland, Hamilton, Ohakea, Napier, Wellington, Christchurch, Dunedian and Invercargill, and (3) forecasts of rainfall in the Waikato, Clutha, Rangitikei, Taupo, Waikaremoana, Mangahao, Cobb, Waitaki and Manapouri catchments. These forecasts are sent to the Island system controllers at Mangakino and Islington and areas used in assessing the demand for electricity during the next 24 hours.

From previous research, it has been established that a 1°C change in temperature causes a change in power demand by about two per cent or 40 to 50 kw of energy. If generated by oil or gas rather than water, this involves additional costs of the order of $ 1,000 for every hour that the extra energy is required. Accordingly, it is clear that the correct prescheduling of power generation (that is, power generation at minimum cost) is related very closely to the accuracy and availability of temperature forecasts. Additional information on the expected rainfall in the key catchment areas of New Zealand is also of considerable importance to the Electricity Department in the day-to-day decision-making involved in the control and use of water storage.

The second type of weather information supplied to the Electricity Department comprises a detailed analysis of the actual temperature departures from normal, at 24 stations in New Zealand at 8 a.m., noon and 5 p.m. for each day. The actual station-hour departures are weighted according to the distribution of the human population, for each Island and for New Zealand as a whole, and the resulting temperature indexes are used by the Electricity Department in monitoring the actual demand for power that occurs. The information is also a vital component in assessing how closely the actual electricity demand, adjusted for temperature condition, meets any government guidelines on national electricity consumption, and also in correctly assessing any additional control which may be required, particularly as may relate to the more than 70 local supply authorities. Work is in progress to incorporate wind and illumination factors into these temperature deviations.

The real value of weather information and weather forecasts for assisting Electricity Department decision-makers to minimise the cost of generating power is difficult to assess; nevertheless, it is clear the weather is a dominant variable affecting domestic power consumption, and that in the absence of the vital weather information cited above, the costs of power generation in New Zealand would be increased by at least $ 1 million a year.

Sheep and Lamb Production and Planning

In a third area of national importance, the sheep-farming industry, the Economic Service of the New Zealand Meat and Wool Boards in association with the New Zealand Meteorological Service, is currently using appropriately weighted weather information (i.e. weighted by the distribution of breeding ewes) to more accurately assess lambing prospects several months before the actual lambing season. This advance information on lambing prospects is of considerable assistance to the Economic Service of the New Zealand Meat and Wool Boards, for the Economic Service is required for planning purposes to predict, a year in advance, the national wool clip per head. To assist in this planning, the Economic Service (Rich and Taylor, 1977) is currently using weighted soil moisture data to more accurately assess the grass available for feed, and hence the sheep live-weight, the live-weight being the major short-term influence on per-head wool production. Work is also in progress in using appropriately weighted weather information for short-term forecasting in the meat industry (Taylor, 1977).

Annex : Weighted Weather Indexes—Interpretation and Value

A key factor in optimising the marketing of meteorological information is the appropriate weighting of weather information, for as is very obvious in applying meteorological information, all things, and all areas are *not* equal (Maunder 1972a, b; Maunder 1977b). Traditionally, however, weather and climate data have in general been confined to point sources, and where area averages, or area descriptions (such as the weather forecast), or maps are given, it is usually assumed that it is on an "equal area" basis. But, "equal area" statements or maps showing last month's weather or tomorrow's weather are of limited use in operational decision-making. For example, a forest industry is basically interested in the weather conditions pertaining to the forested areas, a dairy industry in the weather conditions pertaining to dairying areas, a meat and wool industry in the weather pertaining to the beef cattle and sheep areas, and an electricity industry in the weather pertaining to the (human) populated areas with respect to consumption, but in other respects to energy source areas and transmission routes.

Weather and climatic information must, therefore, be made available in a form appropriate to the user, and to this end weather and climatic indexes weighted according to the distribution of 34 economic and agricultural parameters are currently analysed for 24 areas of New Zealand, including the North and South Islands and New Zealand as a whole. These indexes are : (1) human population, (2) sheep population, (3) dairy cow population; (4) beef cattle population, (5) crop area, (6) land area, (7) area in plantations of timber trees, (8) area in orchards, (9) area in market gardens, (10) area in potatoes, (11) area in wheat, (12) area in vegetable crops, (13) area in seed, hay and silage, (14) area cultivated, (15) lambs shorn, (16) grass land established, (17) grazing land, (18) unused land, (19) total area of holdings, (20) population of breeding ewes, (21) population of calves, (22) number of sheep and hoggets shorn, (23) hydro-electric capacity, (24) hydro-electric capability, (25) fish landed, (26) apple production, (27) pear production, (28) cabbage production, (29) carrot production, (30) lettuce production, (31) onion production, (32) pea production, (33) sweetcorn production, and (34) outdoor tomato production.

These indexes are available for rainfall from January, 1960, for mean temperature from January, 1954, for sunshine from January, 1967, for days of water deficit from January, 1949, and for heating degree days with base 18°, 15° and 12°C, and growing degree days with base 5°, 10° and 15°C, from January, 1950. These new sources of information are available in all cases on a monthly basis, and in most cases also on a partmonth, week, or even daily basis. Of prime importance to their use in operational decision-making is the availability of information in real-time. This real-time availability is now operational, since telegraphic weather data from 50 localities in New Zealand are now used in computer programmes to provide weighted information on a daily basis. In addition, human population weighted data on a selected hourly (and hence daily) basis are available for temperature departures from the average, actual temperatures, and wind speed, from May, 1975.

Apart from their use in operational models—such as those discussed for the dairy industry, the meat and wool industry, and the electricity industry, the weighted values provide significant and informative weather information, in that the information is specifically related to economic/agricultural activities. They provide, in fact, a real alternative to the traditional method of portraying weather and climatic information which treats all areas as being equal. Accordingly, it is no longer necessary or indeed acceptable to say that last month (or last week) temperatures (or rainfall) were above normal. What is required is a statement that last month (or week), temperatures, weighted according to the distribution of a specific population (whether people, animals, crops or their productivity) were above (or below) normal, as shown in Table 13.2. In this way, the vast quantity of weather and climatic information can be both summarised and made operationally useful for many decision-making situations.

TABLE 13.2: New Zealand weighted weather indexes for February, 1976 and 1977

Weighting	Rainfall % of Normal March		Maximum Temperature Departures March		Minimum Temperature Departures March	
	1976	1977	1976	1977	1976	1977
1	2	3	3	4	5	6
Human Population	91	78	−2.2	0.4	−2.3	−0.1
Sheep	125	78	−2.1	0.4	−2.1	−0.2
Dairy Cows	92	58	−2.2	0.4	−2.4	0.1
Beef Cattle	141	74	−2.2	0.3	−2.1	−0.1
Crop Area	107	76	−2.1	0.5	−2.3	−0.4
Land Area	89	79	−1.9	0.5	−2.2	−0.2
Plantations	114	71	−2.1	0.3	−2.2	0.0
Orchards	126	89	−1.7	0.4	−1.9	0.0
Market Gardens	115	67	−2.2	0.4	−2.1	0.0
Potatoes	108	76	−2.4	0.5	−2.3	−0.3

(Contd.)

TABLE 13.2 *(Contd.)*

1	2	3	3	4	5	6
Wheat	103	72	–2.1	0.6	–2.4	–0.5
Vegetable Crops	259	101	–1.9	0.1	–1.3	0.1
Seed, Hay, Silage	101	70	–2.2	0.5	–2.3	–0.1
Area Cultivated	122	75	–2.1	0.4	–2.2	–0.2
Lambs Shorn	151	76	–2.2	0.3	–2.0	–0.1
Grassland Established	123	74	–2.1	0.4	–2.1	–0.1
Grazing Land	66	78	–1.7	0.6	–2.3	–0.3
Unused Land	84	78	–1.7	0.5	–2.3	–0.2
Total Area (Holdings)	98	76	–1.9	0.5	–2.2	–0.2
Breeding Ewes	125	77	–2.1	0.4	–2.1	–0.2
Calves Born	93	59	–2.2	0.4	–2.4	0.1
Sheep/Hoggets Shorn	123	79	–2.1	0.4	–2.1	–0.1
Hydro Capacity	75	70	–1.5	0.8	–2.1	–0.2
Apple Production	125	119	–1.8	0.3	–1.6	0.0
Pear Production	187	106	–1.8	0.3	–1.6	0.0

REFERENCES

Maunder, W. J., 1970 : *The Value of the Weather,* Methuen, London, p. 388.

———, 1971 : "Temperature Forecasts and the Assessment of Electric Power Demand in New Zealand", *N.Z. Meteorological Service Technical Note,* No. 195.

———, 1972a: "The Formulation of Weather Indices for Use in Climatic-Economic Studies : A New Zealand Example", *N.Z. Geographer,* 28, pp. 130-50.

———, 1972b : "National Econoclimatic Models : Problems and Applications", *N.Z. Meteorological Service Technical Note,* No. 208.

———, 1974 : "The Prediction of Monthly Dairy Production in New Zealand Through the Use of Weighted Indices of Water Deficit", *N.Z. Meteorological Service Technical Note,* No. 227.

———, 1977a : "Weather and Climate as Factors in Forecasting National Dairy Production", *Proceedings of the Symposium on the Management of Dynamic Systems, in New Zealand Agriculture,* DSIR, Lower Hutt, June 28, 1977. (in press)

———, 1977b : "National Economic Planning : The Value of Weather Information and the Role of the Meteorologist", *Proceedings of the Ninth N.Z. Geography Conference,* Dunedin, August, 1977.

Rich, M. W. and Taylor, N. W., 1977 : "Econoclimatic Models for Forecasting the National Tailing Percentage and Wool Clip", *N.Z. Agricultural Science,* 11, pp. 65-72.

Taylor, N. W., 1977 : "Short-term Forecasting in the Meat Industry", *Proceedings of the Symposium on the Management of Dynamic Systems in New Zealand Agriculture,* DSIR, Lower Hutt, June 28, 1977.

TABLE 2 (Contd.)

[illegible]	[illegible]	[illegible]	[illegible]	[illegible]	[illegible]	[illegible]
Wheat	104	[illegible]	-1	0.6	2.4	-0.5
[illegible]	149	[illegible]	1.9	0.1	-1.3	0.1
[illegible] Hay/Silage	119	79	2.2	1.5	-2.1	0.4
Area Cultivated	192	75	-2	0.1	2.1	0.2
[illegible]	151	[illegible]	[illegible]	[illegible]	-1.0	-0.1
[illegible]	112	[illegible]	2.1	0.4	[illegible]	-0.1
Grazing Land	66	[illegible]	-1.7	0.6	[illegible]	0.8
[illegible]	[illegible]	[illegible]	[illegible]	[illegible]	2.5	-0.7
[illegible]	98	76	[illegible]	[illegible]	[illegible]	-0.2
[illegible]	[illegible]	[illegible]	[illegible]	[illegible]	[illegible]	-0.4
[illegible]	[illegible]	[illegible]	1.4	0.4	[illegible]	[illegible]
[illegible]	123	[illegible]	[illegible]	0.5	[illegible]	-0.1
[illegible]	[illegible]	[illegible]	[illegible]	[illegible]	[illegible]	0.2
[illegible] Production	[illegible]	[illegible]	[illegible]	[illegible]	[illegible]	0.0
[illegible] Production	[illegible]	[illegible]	[illegible]	[illegible]	[illegible]	[illegible]

REFERENCES

[illegible] 1970. *The Value of the Weather*. [illegible], London, [illegible]

—— 1973. "[illegible] Forecasts and the Assessment of [illegible] Demand [illegible]", *N.Z. Meteorological Service Technical Note* No. [illegible]

—— [illegible]. "The Formulation of Weather Indices for Use in Climatic Economic Studies: A New Zealand Example", *N.Z. Geographer*, [illegible], pp. [illegible]

—— [illegible]. "National [illegible] Models: Problems and Applications", *N.Z. Meteorological Service Technical Note* No. [illegible]

—— 1974. "The Prediction of Monthly Dairy Production in New Zealand Through the Use of Weather Indices and Water Budgets", *N.Z. Meteorological Service Technical Note* No. [illegible]

—— [illegible]. "Weather and [illegible] Factors in Forecasting National Dairy Production", *Proceedings of the Symposium on the Meteorology of Economic Systems in New Zealand Agriculture*, DSIR, Lower Hutt, June 28, 1973 (in press).

—— 1977b. "National Economic Planning: The Value of Weather Information and the Role of the Meteorologist", *Proceedings of the Ninth N.Z. Geography Conference*, Dunedin, August 1977.

[illegible] W. and Taylor, N.W. 1977. "Econoclimatic Models for Forecasting the National Lambing Percentage and Wool Clip", *N.Z. Agricultural Science*, 11 pp. 65-72.

Taylor, N. W. 1977. "Short-term Forecasting in the Meat Industry", *Proceedings of the Symposium on the Meteorology of Economic Systems in New Zealand Agriculture*, DSIR, Lower Hutt, June 28, 1977.

14

AGRICULTURAL METEOROLOGICAL DATA BASE AND AGRICULTURAL PRODUCTIVITY

NOOR MOHAMMAD AND MOONIS RAZA

OF ALL the economic activities of man, agriculture is, perhaps, the one on which environment impinges most, although he often has tried to minimise the restrictions. Nature, with all the diversities in its physical characteristics viz. relief, climate, soil, drainage and vegetation, provides a host of possibilities to man in different parts of the earth surface, and man with his existing level of technological advancement has exploited these opportunities (Hettner, 1947). Hence, from the very beginning the researchers in the field were concerned with the problems of explaining how the variations in environmental phenomena have affected the agricultural scenario (Gregor, 1970). But today when the population is increasing with an alarming rate and the resources are unable to keep pace with it, specially in developing regions, the need to study man's harmony with nature and to understand and coordinate the resources of nature with those of humanbeings is further enhanced. Mc Harg (1966, 1969) has termed it as "ecological determinism" and has explained how nature can impose limits, provide guidelines and assist man in solving environmental problems.

In the early phases of agriculture man was subservient to nature. With his experience of trial and error he used to cultivate crops which suited best to the area according to his perception of environment. He was satisfied with whatever he used to get from agriculture after the damages caused by various calamities. But gradually he tried to break out of the invisible links that had kept him subservient to nature for such a long period of time. He developed interest in understanding the environmental phenomena and also used to be benefited with the age long experience of his predecessors. In India the cultivators, due to ignorance and lack of agricultural meteorological data, used to do agricultural planning according to the local adages regarding environmental conditions which are framed after the experience of ages, and they are true even today in majority of the cases. But these local adages are qualitative in nature and any sound planning cannot be based on them. Therefore, there is a basic need to quantify the environmental attributes precisely on the basis of observations and records in order to meet various human needs by establishing basic guidelines for a balanced ecological planning, modification and adjustment in any area.

For a viable scientific agricultural planning one must understand the basic sets of relationships between the intricately related quadruplet complex i.e. the land, the climate, the soil and the water resources, and agricultural operation, and those between man and available cultivated land. It requires authentic scientific data which could be obtained from agricultural data bases such as agricultural meteorological stations, soil research and testing laboratories, revenue record offices etc. In the present paper only the agricultural meteorological data bases have been taken into consideration.

Among the various components of physical environment, climate plays determining role, directly and indirectly, in shaping the agricultural landscape of an area. While temperature sets the limit for germination of plants and its optimal growth, precipitation is the basic source of moisture. Sunshine, on the one hand, affects the growth and development of plants, while frost, on the other, imposes limits for cultivating a particular crop. Therefore, for the optimal growth and development of agriculture proper information and understanding of various elements of climate is a pre-requisite. If the detailed information regarding the micro-climate of the area is available, the farmers would be in a better position to chalk out a better scientific and rationale agricultural plan which, in turn, will increase the agricultural productivity. The information and knowledge about soil temperature at various depth and in different agricultural seasons, precise forecasting of floods and drought well ahead of time, fortnightly and weekly weather forecasts etc., will be of great help in planning of crop landuse and crop combinations.

The basic source of information and data regarding various climatic phenomena are the records and publications of India Meteorological

Department, which has its own network of observatories all over the territorial limits of the country. Besides its own network of observatories, it also gets data from various observatories partially or fully run and maintained by private, government and quasi-government institutions. But the network of observatories is highly uneven, and it is more so in case of agricultural meteorological stations. These observatories have been set up either arbitrarily or with certain biases and vested interests. Therefore, there are areas which have very high agricultural potential but the network of agricultural observatories is poor, and just opposite is true in case of areas with poor agricultural potential. Consequently, agricultural potential is not properly exploited due to lack of knowledge of micro-agricultural climate. Hence, there is a basic need to study the network of agricultural meteorological stations and the performance of Indian agriculture.

Aim and Objectives

The aim and objectives of this paper are varied and many, which could broadly be summarised as given below :

(i) To map and study the network of agricultural meteorological stations and analyse the regional disparities therein.
(ii) To map and study the regional patterns of Land Resources and Agricultural Productivity.
(iii) To study the locational pattern of agricultural meteorological stations in relation to Land Resources and Agricultural Productivity Regions.
(iv) To find out the association, if there is any, between the network of agricultural meteorological stations and Agricultural Productivity Regions.
(v) To suggest the optimal locations of agricultural meteorological stations with reference to agriculture.

Research Design and Methodology

The regional disparities in the network of agricultural meteorological stations has been examined with reference to : (i) Land Resources Regions, and (ii) Agricultural Productivity Regions.

There are thousands of meteorological stations in India which observe the conditions of various weather phenomena. They differ in nature, quality and character of equipment, number of weather elements to be recorded, recording times, timings and nature of despatch of data to Head Office, nature of maintenance, functions to be performed etc. Therefore, they have been classified into five major categories as given below (I.M.D., 1980):

I Surface Observatories,
II Upper Air Observatories,
III Aeronautical Meteorological Observatories,
IV Climatological Stations for Hydrological Purposes,
V Special Stations.

These observatories are also classified on the basis of their functions, utility and nature of maintenance such as agricultural meteorological stations, seismological stations, marine meteorological stations, flood meteorological offices, etc. But in this paper only Agricultural Meteorological Stations have been taken which observe the conditions of weather elements having agricultural relevance. These stations provide detailed meteorological and biological information and carry research in agricultural meteorology. The meteorological elements observed at these stations include rainfall, air temperature, humidity, soil temperature and moisture, velocity and direction of wind, evaporation, evapotranspiration, sunshine, phenology, incidence and spread of pests and diseases, and the occurrence of weather phenomena like thunderstorm, hailstorm, frost, high winds etc. These observations are recorded at 07.00 and 14.00 hours Local Mean Time. However, rainfall and evaporation observations are recorded at 08.30 hours I.S.T. and 14.00 L.M.T.

Agricultural Meteorological Stations are also classified into following five major categories:

Principal Agricultural Meteorological Station

There is only one such station which is situated at Pune. It provides detailed meteorological and biological information and carries out research in agricultural meteorology.

Ordinary Agricultural Meteorological Stations

These stations provide on a routine basis meteorological and biological information regarding rainfall, air temperature and humidity, soil temperature and moisture, velocity and direction of wind, evaporation, sunshine, and the occurrence of weather phenomena like thunderstorm, frost, high winds, etc. It includes : (i) agricultural meteorological stations which record simple observations on the growth and yield of major crops according to a common standard plan; (ii) agricultural meteorological stations maintained by soil conservation research, demonstration and training centre; (iii) agricultural meteorological stations maintained by agricultural schools, colleges and universities; and (iv) agricultural meteorological stations maintained by agricultural research farms, and allied centres for meeting the meteorological needs for their research programme.

Auxiliary Agricultural Meteorological Stations

These stations provide meteorological and other information such as phenology and the incidence and spread of pests and diseases. The meteorological information generally consists of air temperature, humidity, and rainfall, and may include some of other meteorological parameters included in ordinary meteorological stations. It includes : (i) surface observatories (M.D.) recording phenological observations; (ii) phenological stations for which meteorological data are also recorded by some other Agencies; and (iii) pest and disease reporting stations for which meteorological data are recorded by some other Agency.

Evapotranspiration Stations

At these stations the measurement of evapotranspiration is made by using gravimetric lysimeters in order to assess the water requirement of important crops.

Pilot-Balloon-cum-Micro-Meteorological Stations

These stations (I.M.D.) were started under "Desert Locust Meteorology Scheme". They provide low level winds to assist in the issue of anticipated winds for anti-locust operations.

However, these agricultural meteorological stations have been given judicious scores on the basis of nature and quality of their equipment, number of observation timings, nature and number of transmission of observations to the Head Office, nature of maintenance, and nature and number of weather elements recorded. The following scores have been given to above agricultural meteorological stations :

1. Principal Agricultural Meteorological ....8 scores
2. Ordinary Agricultural Meteorological Stations ....4 scores
3. Auxiliary Agricultural Meteorological Stations ....3 scores
4. Evapotranspiration Stations ....2 scores
5. Pilot Ballon-cum-Micro-meteorological Stations ....1 score

In order to assess the agricultural potential of different areas of the country, Land Resource Regions have been formed on the basis of a study done at Soil Conservation Research, Demonstration and Training Centre, Dehradun (Gupta, et al, 1984). They have divided the country into 20 land resource regions on the basis of relief, vegetation, soil, etc. But in this study these 20 regions have been grouped into following seven major land resource regions on the basis of their similarities and agricultural potentialities :

(i) Very high Potential Regions
(ii) High Potential Regions
(iii) Semi-Medium Potential Regions
(iv) Medium Potential Regions
(v) Low Potential Regions
(vi) Very Low Potential Regions
(vii) Extremely Low Potential Regions

Agricultural productivity regions have been taken from Raza's (1981) study, wherein he has identified and delineated agricultural productivity regions on the basis of productivity per hectare as well as per worker for the years 1960 and 1970 in terms of monetary value. In this paper agricultural productivity regions, formed on the basis of productivity per hectare in the year 1970, has been taken into consideration. These regions are as given below:

(i) High Productivity Regions	> Rs. 1750.00 Per Hectare
(ii) Medium Productivity Regions	Rs. 1100.00-1750.00 "
(iii) Low Productivity Regions	Rs. 750.00-1100.00 "
(iv) Very Low Productivity Regions	< Rs. 750.00 "

Source of Data and Methodology

The study is entirely based on secondary data and information. The detailed information and data regarding the location, nature function and characteristics of agricultural meteorological stations have been taken from India Meteorology Department, Mausum Bhawan, Lodhi Road, New Delhi, and its publication — *Observational Organisation as on Ist January*, 1980 (I.M.D.). The Land Resource Regions and Agricultural Productivity Regions have been taken from the studies of Gupta et al, (1984) and Raza (1981).

Cartographic and statistical techniques have been used in this paper. First of all a map showing location of all classes of agricultural meteorological stations with their total scores has been prepared (Fig. 14.1). Another map showing the agricultural resource regions together with their ranks is also drawn on a tracing sheet (Fig. 14.2), which has been superimposed on first map. The scores of all agricultural meteorological stations falling in each land resource region have been added together which gives its strength in terms of network of agricultural meteorological station. Percentage strength of each region with respect to total scores of observatories in India and its percentage contribution to the total area of the country has also been worked out. Finally, in order to test this relationship between land resource region and network of agricultural meteorological stations, correlation co-efficient is calculated. T-test has also been applied in order to assess the level of significant of the association. Land Resource Regions have also been ranked in order of their agricultural potentiality, and so also the network of agricultural meteorological stations according to their percentage strength with reference to total meteorological scores in order to test the rank correlations between them. Students test (t) is also applied to it.

In order to probe further, a map showing agricultural productivity regions is drawn on a tracing paper (Fig. 14.3), and it has also been superimposed on first map. The scores of agricultural meteorological stations falling in each agricultural productivity region have been counted and its percentage contribution to total scores of stations is calculated. The percentage share of net sown area of each region to the total net sown area of the country is also taken into consideration. Finally, correlation co-efficient between these two variables has been calculated. The agricultural productivity regions have to be marked productivity-wise as well as according to their percentage strength with respect to the scores of agricultural meteorological stations. Finally rank correlations has been worked out. T-test is also applied in order to know the level of significance.

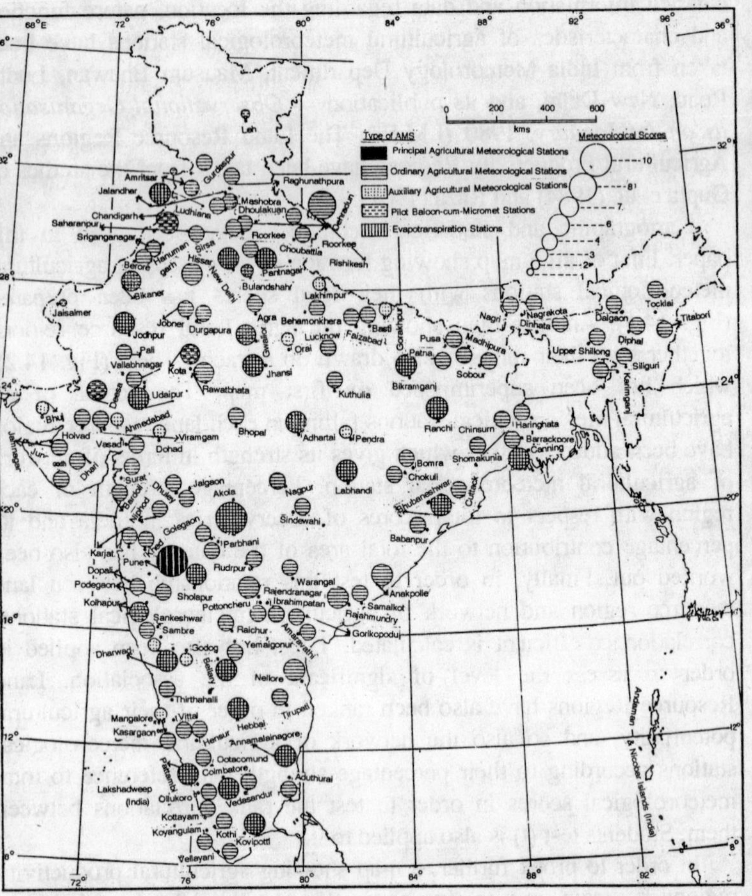

FIG. 14.1: Agricultural meteorological data base.

Land Resource Regions and Agricultural Meteorological Data Base

The physical environment of man is made up of landforms, climate, soil, water, minerals, natural vegetation and native animal life. This resource complex (or land as economist call it) provides man with living space, food and a vast variety of raw materials used for the satisfaction of human wants. Land resource plays a strategic role in determination of man's economic, social and cultural progress which could be easily understood through reference to economic history of different countries. The localised nature of land resources give rise to a keen struggle for political control of raw materials by the old colonial powers and shaped considerably their political and economic histories.

Soil Conservation Research Demonstration and Training Centre, Dehradun has prepared the land resource map of India by integrating the soil map of India (Raychaudhuri, 1963), Forest map of India (Anonymous, 1968), Forest types of India (Champion and Seth, 1968), rainfall, landuse, water resource and elevation map of India (Chatterjee, 1957). They have identified 20 land resource regions and 186 land resource areas but in this paper these 20 regions have been grouped into 7 land resource regions on the basis of their similarities and agricultural potential (Table 14.1).

The distribution of Agricultural Meteorological Stations (hereafter written as AMS) according to land resource region is given in Table 14.2. The region of very high agricultural potential extends from Pakistan border in the northwest to the border of Bangladesh in southeast. It covers almost the core of the Indo-Gangetic plain in

TABLE 14.1: Land resource regions

Region	Rank	Area in km^2	%age to total area
Very High Potential Region	I	472,000	14.43
High Potential Region	II	154,500	4.78
Semi-Medium Potential Region	III	921,250	28.37
Medium Potential Region	IV	265,520	8.20
Low Potential Region	V	789,250	24.31
Very Low Potential Region	VI	346,000	10.66
Extremely Low Potential Region	VII	300,250	9.25
Total		3,248,770	100.00

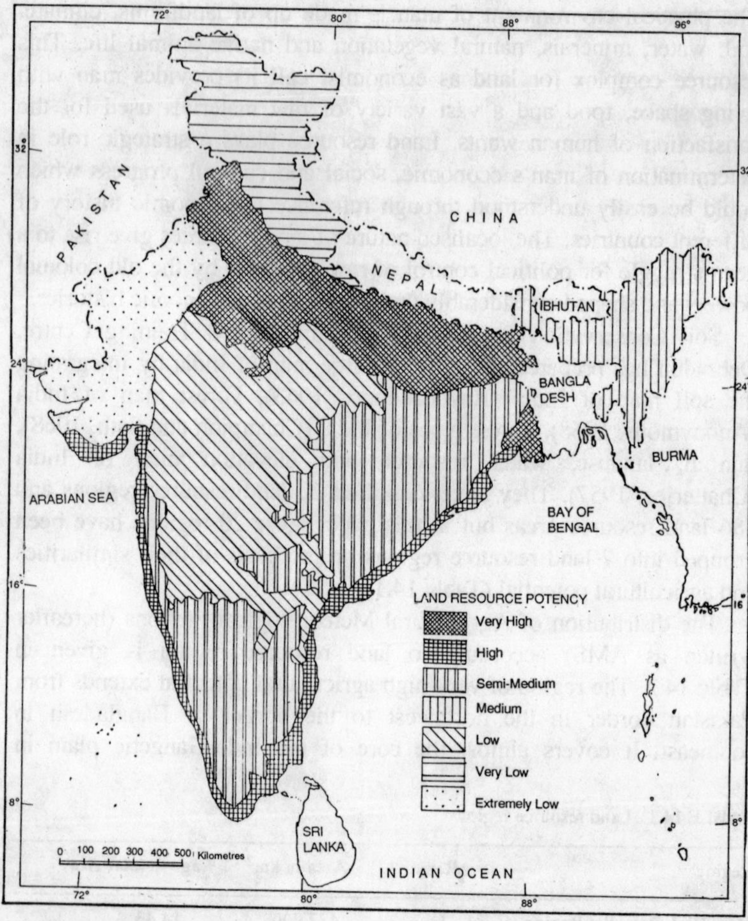

FIG. 14.2: Land resource regions.

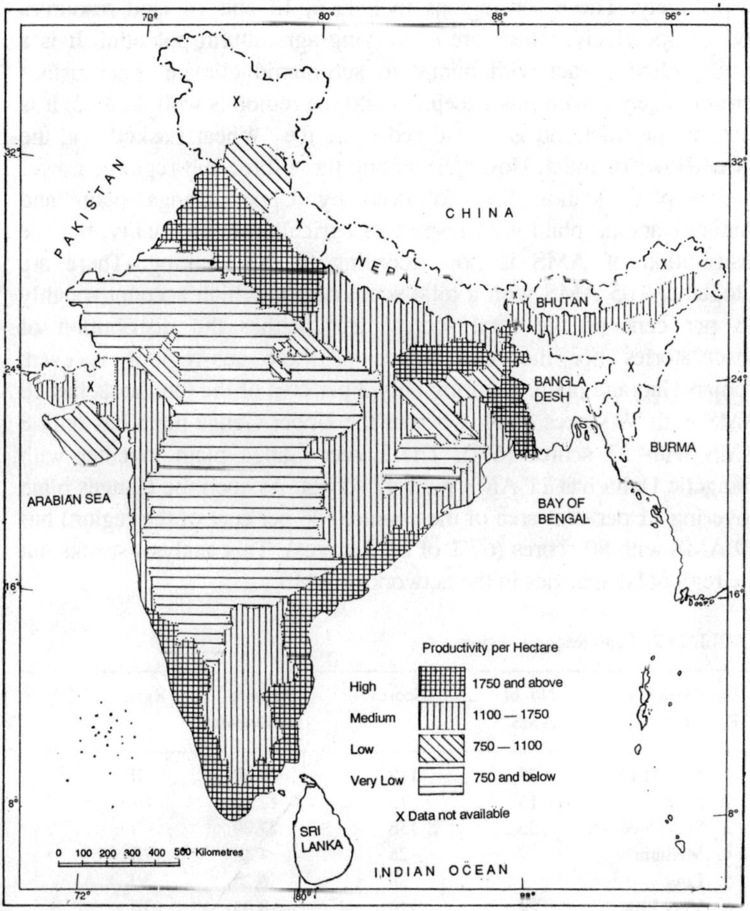

FIG. 14.3: Agricultural productivity regions.

India. It includes Punjab-Haryana alluvial plain, Upper Ganga plain and Lower Ganga plain including Gangetic Delta region, which cover 3.10, 6.15 and 5.18 per cent of the total area of the country respectively. These sub-regions include 6, 13 and 14 land resources areas respectively, which are of varying agricultural potential. It is a fertile alluvial tract with humid to sub-humid climate. The surface terrain is gentle with few exception and the region is well drained. It is most fertile tract and is considered to be the "Wheat Basket" and the "Rice Bowl"of India. However, among these three sub-regions, Lower Ganga plain stands first followed by Upper Ganga plain and Punjab-Haryana plain with respect of agricultural potentiality, but the distribution of AMS is not according to this ranking. There are altogether 105 AMS with a total score of 108 which account roughly 18 per cent of the total scores. But neither the distribution of observatories nor the total scores of these sub-regions is even. Punjab-Haryana plain, covering only 3 per cent of the total area, has 10 AMS with 39 scores (8%), whereas the Upper Ganga plain has only 8 AMS with 33 scores (6%). The Lower Ganga plain together with Gangetic Delta has 11 AMS with 47 scores. As such the Ganges plain covering 11 per cent area of the country (86 per cent of the region) has 19 AMS with 80 scores (67% of total scores). This analysis speaks out the regional disparities in the network of AMS.

TABLE 14.2: Land resource region

Land Potential Regions	No. of AMS	Scores	%age to total scores	Rank
1. Very High	28	119	20.48	II
2. High	16	73	12.56	IV
3. Semi-medium	33	158	27.20	I
4. Medium	7	26	4.48	VII
5. Low	9	36	6.20	VI
6. Very Low	24	110	18.93	III
7. Extremely Low	14	59	10.15	V
Total	131	581	100.00	

AMS = Agricultural Meteorological Stations

The region of high agricultural potential is confined to the coastal areas leaving the Assam and Gujarat coastal belts. It includes the eastern and western coastal regions which cover nearly 1.88 and 2.90

per cent of the total area of the country. The western coastal region includes 7 land resource areas, whereas the eastern coastal region has 8 land resource areas. It is a fertile belt with fine alluvial and coastal soils. The climate is, by and large, humid to sub-humid, but in certain areas semi-arid conditions also prevail.

This region has good network of AMS partially due to its high agricultural potential and partially due to its strategic location. It occupies forth rank in terms of meteorological scores, although it is at the rock bottom in terms of areal strength. There are 16 AMS with 73 scores which accounts roughly 13 per cent of the total scores. But the distribution of the stations and the scores is neither uniform nor in accordance with the agricultural potentiality of the region. The eastern coastal region which not only covers the larger area but also has more agricultural potential, has fewer number of AMS as compared to the western coastal region. The western coastal region has 11 AMS (more than double of the eastern coastal region) with 47 scores which is roughly 65 per cent of the total scores in the coastal region; although it covers only 37 per cent of its area. However, in general the region enjoys fourth rank in terms of percentage strength of scores. A closer scrutiny of the fact at land resource area level shows further disparities in the network of AMS. Majority of the agriculturally more potential areas have fewer number of AMS which manifest that the network of AMS is not on the basis of their regional requirements. They are either arbitrarily located or due to certain political and vested biases.

The region of semi-medium agricultural potential covers the highest percentage area of the country and is found in its south and southeastern portion. It covers about 28 per cent of the total area of the nation. There are two sub-regions in it viz. the eastern red soil region; and the southern red soil region, which cover 17.65 and 10.72 per cent of the total area of the country respectively. There are 17 land resource areas in the former, whereas the latter contains 25 land resource areas. Variety of soils with differential rate of fertility are found in these two sub-regions. The region enjoys humid to dry climatic conditions.

There are 33 AMS with 168 scores (29% of the total scores) in this region. But their spatial distribution is neither in consonance with their areal coverage nor in terms of agricultural potentiality. The eastern red soil region, covering 62 per cent of the total area of this region, has 14 AMS with 62 scores, while the southern red soil region, covering only 38 per cent area, has 19 AMS with 93 scores. It shows that about 58 per cent AMS and 55 scores are in southern red soil region, which is agriculturally less potential. Similar results emerge when we analyse at land resource area level. The land resource areas with high agricultural potential, have, by and large, lesser number of AMS with fewer functions resulting to low scores.

The whole of eastern India starting from West Bengal-Assam border has been included in the medium potential region, although there are a lot of variations in terms of land potential. The fertile-alluvial valley of Brahmaputra river is on the one hand, while the north eastern Himalayan alpine grasses and meadows on the other, but still there are similarities in several physical parameters. Hence, they are put into one land resource region viz. northeastern region which covers around 8 per cent of the area of the country. There are three distinct sub-regions in it viz. northeastern Himalayan alpine grass and meadows region; northeastern forest region; and Assam valley region, which contribute 0.50, 4.97 and 2.73 per cent of the total area of the country respectively. There are 20 land resource areas with varying agricultural potentials in this region. It has humid climate with thick cover of natural vegetation. It contains variety of soils ranging from alluvial soil of Brahmaputra valley to the mountain soil in the upper regions.

The region is not properly served with AMS network. There are only 7 AMS in this region with 26 scores which is only 4.5 per cent of the total. Regional disparities could also be clearly observed in this region. The northeastern alpine and meadows region does not have any AMS, while Assam valleys have 3 AMS with 12 scores and northeastern forest region has 4 AMS with 14 scores. In fact, Assam valleys have far more agricultural potential as compared to the forest region. Hence, it requires more AMS. Therefore, the AMS should be set up in different parts of this region so that agriculture may further develop.

The region of low agricultural potential is located in the centre of the country covering northern part of Peninsular India, Vindhyan region, and southeastern flank of the Aravalis. It is second largest land resource region in terms of area covering more than 24 per cent of the total area of the country. There are two sub-regions in it viz. mixed yellow-red and black soil region; and black soil region, which cover 3.50 and 20.73 per cent of the total area of the country. The former sub-region is found in two patches: One is located on the southeastern flank of Aravalis, while the other is found in the northeastern part of the Vindhyan. This region has lower agricultural potential as compared to the latter.

There are 24 AMS with 110 scores. It has roughly one-fifth of the total scores, although it accounts nearly one-fourth of the total area of the country. Out of this the black soil region, covering 85 per cent area of this region, has 22 AMS (92% of the total) with 100 scores (91% of the total) which shows that in this region there is association between areal strength of the region and network of AMS at meso level. But the closer scrutiny of the distribution of these two parameters does not prove this fact with positive strength. Therefore, it is suggested that more AMS should be instituted according to the land resource potential of the area.

The region of very low agricultural potential is found in the northern Himalayan region covering the states of Jammu and Kashmir, Himachal Pradesh, and northern parts of Punjab and Uttar Pradesh. It covers more than one-tenth of the total area of the country and consists of three sub-regions viz. northern Himalayan snow clad region; northern Himalayan Alpine grasses and meadows region; and northern Himalayan forest region, which cover 3.56, 3.02 and 4.08 per cent area of the country respectively. This is mountainous region mostly covered with snow, meadows, grasses and forests. Only the river valleys have high agricultural potential where there is alluvial and mountain soil. There are altogether 25 land resource areas of varying agricultural potential in this region. It is, by and large, fit for horticulture which requires more detailed information regarding the conditions of various elements of climate.

There is only one AMS with 2 scores in the snow clad region, whereas there is none in the alpine grass and meadows region. But in the northern Himalayan forest region, there are 8 AMS with 34 scores which is, in fact, around 6 per cent of the total. These AMS are located in the river valleys which are rich in agricultural potential.

The extremely low potential region lies in the northwestern part of the country. It covers slightly less than one-tenth area of the nation. It consists of Rajasthan desert, Runn of Kutch and Gujarat alluvial plain. These sub-regions cover 5.87, 1.44 and 1.94 per cent of the total area of the country respectively. With the exception of Gujarat alluvial plain, the region is either desert with infertile sandy and desert soil or marshy. Both these regions have low agricultural potential. However, Gujarat alluvial plain has high agricultural potential. The northern part of the desert region has also reasonable agricultural potential.

However, this region has a good net work of AMS inspite of low agricultural productivity. The large number of AMS in this region is due to its location in the border area. There are 14 AMS with 59 meteorological scores. The lion's share of this score (54%) goes to the desert region, while 34 per cent is in Gujarat alluvial plain. The remaining share goes to Runn of Kutch.

The above analysis of spatial distribution of agricultural data base in relation to land resource regions clearly indicates that the network of AMS is not in consonance with land resource regions. The highly agricultural potential regions do not have even their proper share in AMS, although they should have more. Secondly, the number of AMS is also not adequate at national level. Thirdly, the regional disparity in the network of AMS goes on increasing as one moves from macro to micro level.

However, the statistical technique applied to test the relationship between AMS network and land resource regions at macro level gives positive result. The correlation co-efficient, applied to the percentage areal strength of each land resource region, and the percentage strength of each region with respect to scores, gives a positive result of 0.41. Although the result is positive, it does not show very healthy relationship which has also emerged from our descriptive analysis. When '*t*' test has been applied to test it further, the level of significance comes to be nil which further strengthens our hypothesis viz. there is a regional disparity in the network of AMS. Rank correlationship has also been worked out at micro level for these two parameters. The '*r*' value comes to be 0.51 which is significant at 0.50 per cent level when tested with students test.

Agricultural Meteorological Data Base and Agricultural Productivity

Agricultural productivity is the indicator of the level of exploitation of agricultural land potential with the help of level of technology at a given point of time. It also manifests socio-economic infrastructure and the physical complex of the area which could be developed and properly utilised if detailed information regarding various elements of climate is available. Therefore, it has been endeavoured to find out the correlation, if there is any, between agricultural meteorological data base and agricultural productivity.

The concept of agricultural productivity is highly debatable and various studies have been done in this connection all over the world

(Kendall, 1939; Stamp, 1960; Shafi, 1976; Enyedi, 1964; and Mohammad and Singh, 1981), but in this paper "Agricultural Productivity Regions", worked out and delineated by Raza (1981) for the year 1970 have been taken. National Sample Survey (NSS) regions have been taken as unit of study. The production of various crops (Area × yield) has been multiplied with constant national level price in order to get it in monetary value. The production per hecater is obtained by dividing the total production with total cropped area in each NSS region. Finally, on the basis of this result, agricultural productivity regions, as give in Table 14.3, are formed. Table 14.4 shows the details of network of AMS with reference to agricultural productivity regions. The high productivity region covers nearly one-fifth of the total net sown area of the country and has three definite cores viz. Punjab-Haryana and Western Uttar Pradesh; coastal plains of Tamil Nadu, Andhra, Kerala and Karnataka; and Deltaic West Bengal. This region has good network of AMS. There are 41 AMS (around one-third of the total AMS) with 190 scores (34% of the total). Out of this 15 AMS with 67 scores and 19 AMS with 91 scores are in the former two regions respectively, while the remaining 4 AMS with 32 scores are in the third region. There is viscious circle in the high productivity region and AMS network. These regions were productive, so more AMS have been established there. Since more AMS are there providing more agricultural meteorological information, they become more productive. Therefore, the productive regions kept on becoming more and more productive, whereas the less productive regions could not keep pace with them. Besides this, the reason for better AMS network in this region is its strategic location. The former region is located near India-Pakistan border, while the latter two regions have coastal location.

TABLE 14.3: Agricultural productivity regions

Region	Rank	Productivity Ranges (Rs.)	%age of NSA to total NSA*
High Productivity	I	> 1750	19.75
Medium Productivity	II	1100 – 1750	29.15
Low Productivity	III	750 – 1100	16.38
Very Low Productivity	IV	< 750	34.72
		Total	100.00

* NSA = Net Sown Area.

TABLE 14.4: AMS network and agricultural productivity

Productivity Regions	No. of AMS	Score	%age to total scores	Rank
High Productivity	41	190	33.68	I
Medium Productivity	44	175	31.03	II
Low Productivity	11	47	8.34	IV
Very Low Productivity	32	152	26.95	III
Total	128	564	100.00	

The medium productivity region covers nearly 30 per cent of the net sown area of the country. It almost skirts the region of high productivity. Three well marked areas of medium productivity are easily identified on the map viz. South Deccan; Middle Ganga plains together with Mahanadi basin and Brahmaputra valley; and Konkan region of western coast and Chattisgarh plains. Besides these, a small region has also emerged in the Runn of Kutch having only one AMS with 3 scores. However, there are altogether 44 AMS with 175 scores accounting more than 31 per cent of the total scores. Out of this 26 stations with nearly 60 per cent scores of this region are located in the alluvial plains of Ganga, Mahanadi, and Brahmaputra, whereas 9 AMS with one-fifth of total scores of the region are located in the Konkan region and Chattisgarh plains. The South Deccan has 8 AMS with 19 per cent scores of this region. This analysis throws sufficient light on the regional disparities in the AMS network with respect to agricultural productivity.

The low productivity region covering around 10 per cent of the total net sown area is the smallest region scattered almost in the northern part of the country. The region, as a whole, has 11 AMS with 47 scores which accounts more than 8 per cent of the total. There are five sub-regions of almost equal size but the density of AMS network is not the same in all the regions. The northeastern Punjab (Hilly areas) and Himachal Pradesh have 2 AMS with 8 scores, while the Bundelkhand region of Uttar Pradesh has 3 AMS with 11 scores. There are 3 AMS with 12 scores in Gujarat region, while the southwestern Aravalis region has 2 AMS with 10 scores. The Rhur cluster has only one AMS with 6 scores.

The region of very low productivity covers about 35 per cent of the total net sown area of the nation. An island of very low productivity is found in the central Orissa which has only 2 AMS with 8 scores. The biggest chunk of this region constitutes the Hungry Belly of India and is located in Central India extending from the desert region of Rajasthan in the north to the Mysore plateau in the south. This sub-region has 30 AMS with 144 scores which is more than one-fourth of the total. Unfortunately, this region has altogether 25 per cent AMS with nearly 27 per cent scores, although it covers 35 per cent of the total net sown area. Therefore, it requires more AMS so that agricultural potential may properly be developed.

The above analysis further proves our basic hypothesis of regional disparities in the network of AMS. The percentage strength of meteorological scores is in accordance with the productivity rank of the regions except in case of low and very low productivity regions where it is interchanged. Similar is the case with degree of concentration of scores in the high productivity region. Each per cent of net sown area has 9.62 scores, whereas the very low productivity region has 4.38. The figure for medium productivity region is 6.00, whereas it is just 2.86 in case of low productivity region. It shows that the prosperous regions, due to their better socio-economic conditions resulting to more political pull and push, have more AMS, which leads to more agricultural productivity. And more agricultural productivity leads to the establishment of further more AMS. As such this vicious circle continues and productive regions become more productive and low productive regions further deteriorate leading to more regional disparities.

The statistical technique (correlation co-efficient) applied to test, their relationship gives almost similar results. The 'r' value is 0.49 which shows the positive relationship but the tie is not very strong which also emerges from the result of 't' test. It is 0.79 which indicates the significance level as nil. In order to probe further Rank Correlation Co-efficient has also been worked out and the result is 0.8 which is a high positive value indicating better relationship. Students test (t) has also been applied which gives 50 per cent significance level.

Conclusion

The above analysis leads to the following specific conclusions :

1. The network of AMS, in general, is very poor in our country, which being basically an agricultural country, requires detailed

and timely information regarding various elements of climate for better agricultural planning leading to more agricultural production.

2. Whatever AMS are available, there is considerable regional disparity in their spatial network. There are certain areas having large number of AMS, while there are others who does not have even single AMS.
3. The location of AMS is arbitrary and is, by and large, influenced by political and vested biases leading to regional disparities.
4. There is a lot of regional disparities in the land resource potentials in our country which is bound to be there due to varied environmental conditions and complex physical framework.
5. Although there is a positive relationship between network of AMS and land resource potentials at meso level but the relationship is not strong as is also evident from the results of 't' test. The association further deteriorates if examined at micro level i.e. 'land resource area' level. The study reveals that highly land resource potential areas have fewer number of AMS with little scores and vice-versa. It leads to more and more exploitation of land resources of one region and neglecting the other which could also contribute to the agricultural production of the country.
6. There are regional imbalances in agricultural productivity which is leading or may lead in the future to socio-economic conflicts. With all our planning which claims to be oriented towards social justice, the gap between more productive and less productive regions keeps on widening.
7. The network of AMS is little more oriented towards the agricultural productivity region. The 'r' values in case of correlation co-efficient and rank correlation are 0.49 and 0.80 respectively which shows positive relationship at 50 per cent significant level.
8. Although this relationship is little stronger, it does not, perhaps, mean that more AMS with higher scores have lead to higher agricultural production. In fact these areas were productive even earlier leading to socio-economic prosperity of the inhabitants which has lead to have, through political push and pull, more and more AMS. This process has further increased

agricultural productivity of the region. As a result regional disparity in agricultural productivity kept on increasing.

9. It is, therefore, suggested that a detailed scientific survey of the land resource potentials of the country should be done and the areas of different agricultural potentials must be identified and mapped. Besides, the suitability of these land resource potential regions with respect of crops and cropping pattern should also be worked out. Keeping the land capability and suitability in view, adequate number of AMS, with the equipments observing all those elements of climate, which affect the crops and cropping system, should be located in these regions.

REFERENCES

Anonymous, 1968 : *Forest Map of India,* Forest Research Institute and College, Dehradun.

Champion, H. G., and Seth, S. K., 1968 : *Revised Survey of Forest Types of India.*

Chatterjee, S. P., 1957 : *National Atlas of India,* National Atlas Organisation, Govt. of India, Calcutta.

Enyedi, G. Y., 1964 : Geographical Types of Agriculture, in Pecsi, M. (Ed.) *Applied Geography in Hungry.*

Gregor, H. F., 1970 : *Geography of Agriculture : Theme in Researches,* p. 31.

Gupta, et. al, 1984 : Land Resource Regions and Areas of India in Ramesh, A. (Ed), *Resource Geography,* Heritage Publisher, New Delhi, p. 24-34.

Hettner, A., 1947 : *Allgeiunen Geographic Der Menchen* (General Geography of Humanity), Vol. I, p. 1.

Kendall, M. G., 1939 : 'The Geographical Distribution of Crop Productivity in England', *Journal of Royal Statistical Society,* 102, pp. 21-48.

Mc Harg, I. L., 1966 : 'Ecological Determinism in Darling M. F. and Milton J. (Eds), *Future Environment of North American,* p. 190.

———, 1969 : *Design with Nature,* National History Press, New York, p. 190.

Mohammad, N. and Singh, R., 1981 : 'Measurement of Crop Productivity : A Review', in Mohammad, N.(Ed), *Perspectives in Agricultural Geography,* Vol. IV, Concept Pub. Co., New Delhi.

Mohammad, N. (Ed), 1981 : *Perspectives in Agricultural Geography,* Vol. IV, Concept Pub. Co., New Delhi.

Raza, M., 1981 : 'Regional Disparities in India : A Preliminary Exploration of the Regional Dimensions of Agricultural Development', in Mohammad, N. (Ed), *Perspectives in Agricultural Geography,* Concept Pub. Co., New Delhi, pp. 103-44.

Raychaudhuri, S. P., 1963 : *Land Resources of India : Soils, their Classifications, Occupations, Occurrence and Properties,* Planning Commission, New Delhi.

Stamp, L. D., 1960 : *Our Developing World,* London, pp. 108-09.

Shafi, M., 1976 : 'Approaches to the Measurement of Agricultural Typology', in Misra, V. C., Ayyar, N. P. and Kumar (Eds), *Essays in Applied Geography,* pp. 67-77.

15

ECOLOGY AND RICE CULTIVATION AMONG THE APA TANIS OF ARUNACHAL PRADESH

ARUN KUMAR SINGH

IT IS a general feeling that mountains are inaccessible and not suited for better human habitation. This assumption is based on a fact that most of the human civilisations have developed in the river plains or very near to the foothills. Mountains always remained a sparsely populated region in the world. However, since time memorial, man has utilised the mountain ecosystem for subsistence activities on a marginal level such as nomadic herding with emphasis on seasonal migration. Keeping this contex in view the present paper tries to focus its attention towards a society which has utilised the mountain ecosystem for cultivation of rice in its well organised wet fields. The society is that of the Apa Tanis, a small tribe (population 18,000 in 1981 Census approximate) of Lower Subansiri District of Arunachal Pradesh (Fig. 15.1). Efforts have also been made to understand the dynamics of adaptation of the Apa Tanis with reference to mountain ecosystem and rice cultivation.

The Society

Apa Tanis society differs from the surrounding societies of Subansiri region. Enclosed from all sides by high hills and hostile climate the Apa Tanis have developed their resources in such a way that the whole society has become self-sufficient. Seclusion and self-sufficiency are the key-notes of Apa Tanis Society (Furer-Haimendorf, 1962). In the absence of any archaeological data it is difficult to ascertain from where and when they have migrated and settled in this valley. But

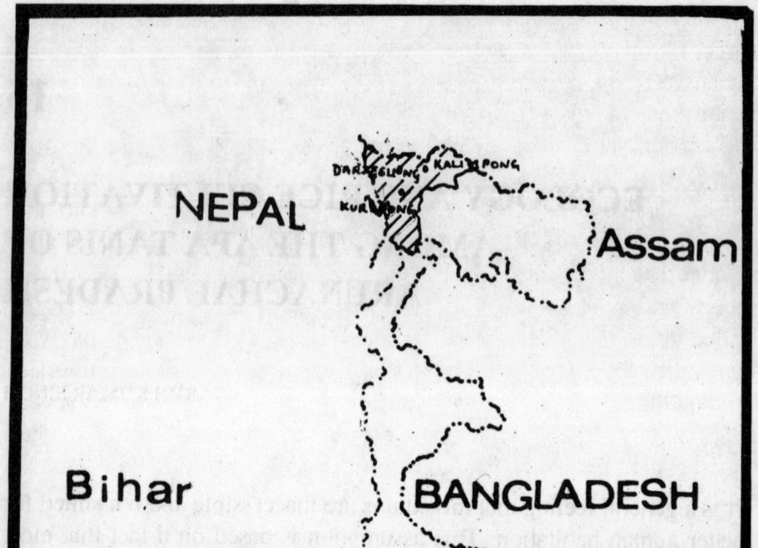

FIG. 15.1: Location map district Darjeeling.

judging from the way in which they have transformed their environment, one can safely assume that many centuries must have passed since the forefathers of the present population first set foot in the valley. A tradition current among the Apa Tanis tells that their ancestors came from a country to the north or north-east situated near two rivers known as Supupad-Pudpumi. These names may refer to two tributaries of the Subansiri, but neither the Apa Tanis nor anyone else is likely to identify this legendary country of origin. All Apa Tanis agree, however, that at one stage in their migrations they crossed the Subansiri River from north to south (Furer-Haimendorf, 1962). From racial point of view the Apa Tanis belong to Palaeo-Mongoloid type and are similar to their neighbouring tribes—the Nishis and the Hill Miris. Apa Tanis speak their own dialect which is nearer to Dafla (Nishi) language of North Assam group which again falls into the Tibto-Burman sub-family or Tibto-Chinese family. Apa Tanis also believe that in course of their migration they were divided into three groups and reached the present place of dwelling. The first group settled in the Hong Village which is also the biggest village of the Apa Tanis, the second group settled in Hari/Bela and the third group settled in Haja/Michi Bamin/Duta villages (Fig. 15.2). There are the seven oldest villages of the Apa Tanis. Three villages are situated on the eastern side of Kale river whereas four villages are settled on the western side. At present there are 23 villages of the Apa Tanis.

The Region

The name Lower Subansiri District in which lies the Apa Tani's plateau, is derived from the River Subansiri which originates in the Himalayas, flows southwards and meets the mighty Bhramputra River near Majuli (Fig. 15.1). River Subansiri also forms to some extent the western boundary of the district. The whole district is covered with deep dense forest with high mountain ranges. The ground rises in tier after tier of mountain spurs, forming deep gorge-like valleys down which descend numerous mountain torrents and rivers, from the thickly forested foot-hills where the vegetation is tropical to the dizzy heights of the snow line in the extreme north (Chowdhury, 1970). The Apa Tani's plateau lies approximately in the middle of Lower Subansiri District. It is well connected with Itanagar—the State capital of Arunachal Pradesh as well as with North Lakhimpur in Assam by road. However, Ziro, the District headquarters of Lower Subansiri has an airfield. Apart from Ziro being well connected with rest of the state and country, all the villages of the Apa Tanis are also well connected with all weather roads.

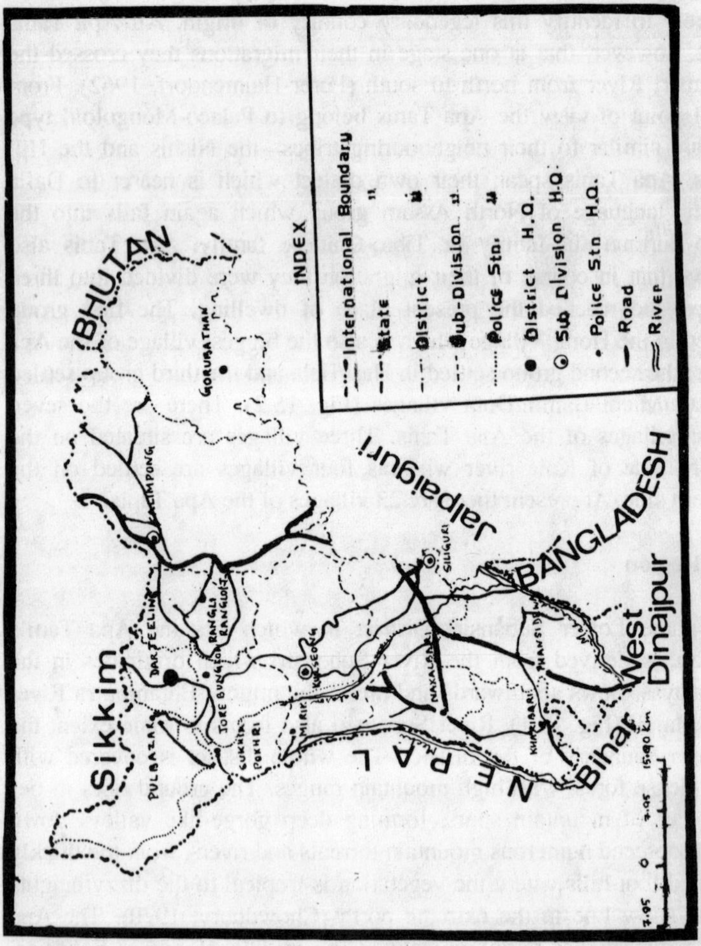

FIG. 15.2: Darjeeling distrist.

During the British period the whole of NEFA (North Eastern Frontier Agency) was a part of Assam as a Non-Regulation Province. In 1873 Inner Line Regulation was extended to the Hill area of Assam. Beyond the Inner Line no British subject of certain classes or foreign residents could pass without a Licence. With a view to collect more knowledge regarding the economics and culture of the hill people a post of Political Officer with Headquarters at Saidya was created in 1881-82. In the year 1912-13 a separate district was formed as Subansiri and was brought under the control of the Political Officer. In 1914-15 the Lakhimpur Frontier Tract was reconstituted and Subansiri fell under Western section with Political Officer stationed at Charduar. In 1919 the Western section was named as Baliapara Frontier Tract with its Headquarters at Charduar. In 1945 Baliapara Frontier Tract was divided into two-Se-la sub-Agency and Subansiri Area. In 1946 Subansiri area was named as Subansiri Frontier Division. The Political Officer of Subansiri Area had his Headquarters first at North Lakhimpur with a base camp at Kimin. The Headquarters were shifted to Kimin from North Lakhimpur in 1950 and finally to Ziro in 1952. In 1965 according to the North East Frontier Agency (Administration) Regulation of 1965 Subansiri Frontier Division came to be known as Subansiri District and Political Officer was redesignated as Deputy Commissioner. In 1980 Subansiri District was divided into two districts and were named as Lower Subansiri and Upper Subansiri. The Apa Tanis plateau falls under Lower Subansiri District with Headquarters at Ziro. In brief, we have presented the history of administrative development of Subansiri region because it had laid a great impact on the development process of the region in general and Apa Tanis in particular.

Evolution of Rice Cultivation among the Apa Tanis

The corner stone of Apa Tanis agriculture is the cultivation of rice on irrigated terraces. Rice is a staple food, and all other crops are grown mainly to provide variety of diet and to utilise those portions of the country not suitable for irrigation. Rice too is the principal item in the Apa Tanis export (Furer-Haimendorf, 1946). Rice is not only the principal crop of the Apa Tanis but the whole economy, social and ritual pattern revolves round rice cultivation. But the question of origin of Apa Tanis rice cultivation remains unanswered by the early writers (Furer-Haimendorf, 1946, 1955, 1962, 1980); (Bower, Ursula Graham,

1953), (Elwin, Verrier, 1957). From the interviews with many Apa Tanis it was found that during their course of migration they did not bring rice with them. It seems that they brought millet (*Eleusine Coracana*), maize, pine (*Pinus excelsa*) and bamboo. These crops and trees vary from the crops and trees cultivated and grown from the neighbouring tribes Daflas (Nishis) and Miris. However, it can be assumed that Apa Tanis settled down in a valley which was a marshy land (Izzard, R, 1951) and rice must be growing it its wild form. Apa Tanis must have developed the present rice variety from the local wild rice during their long occupation of the valley. Looking at the geographical situation of the valley it can also be presumed that there were no option left for the early Apa Tanis settlers but to develop rice crop. The millet and maize which were brought by them from the place of migration was not suited to the marshy land. This belief that rice is a local crop gets more confirmed if we look at the map prepared by Tadaya, Watabe (1985) to demonstrate the geographical origin of cultivated rice (Fig. 15.3). He confirms that rice cultivation originated in the Assam-Yunnam Zone and spread to the various countries of South East Asia, China and Japan (Fig. 15.4). It is also found that nearly all the tribal population of North Eastern States in general and Apa Tanis neighbouring tribe—the Dafla (Nishis) (Shukla, 1959) and the Hill Miris (Chowdhury, 1970) grow rice in their Jhum fields. Looking at the rice cultivation system of the Apa Tanis one can easily feel that rice cultivation is being practiced by them since time memorial. The ecological conditions also help and support rice cultivation and Apa Tanis have excelled in it.

Ecology and Rice Cultivation

Rice, as we have noted earlier the most distinctive crop grown by the Apa Tanis, is cultivated intensively but to a great extent on subsistence level in the fragmented but well maintained fields. On the success or failure of rice depends the whole economy of the Apa Tanis. If we observe which shows the agricultural landuse of the Apa Tanis it becomes clear that above 90 per cent of the arable land is under rice cultivation. Whenever land is suitable and water is available rice is being grown.

Among the ecological factors effecting rice cultivation, availability of water is most important. The field has to be submerged under approximately 6" of fresh and slowly moving water for at least 75

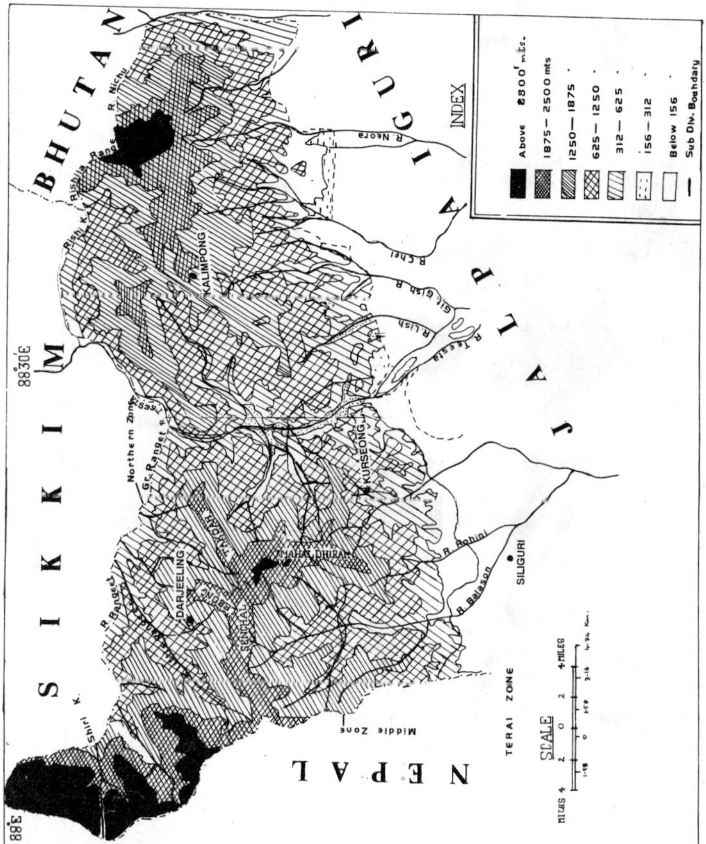

FIG. 15.3: Origin of rice in Apa Tanis.

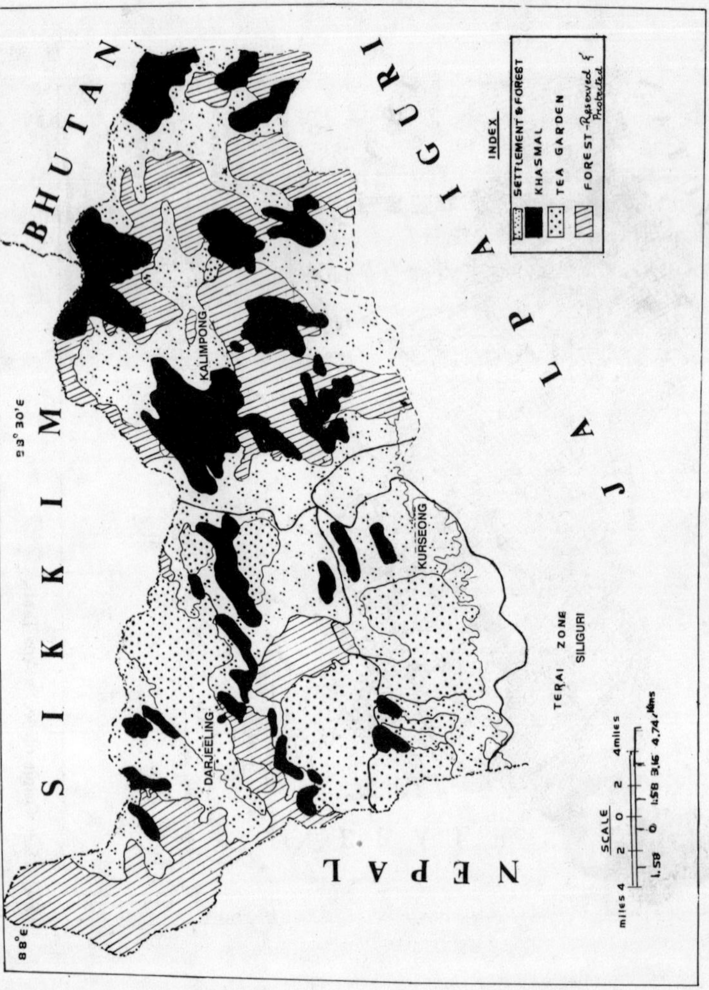

FIG. 15.4: Landuse.

days. If rainfall is not regular fields have to be properly irrigated. Keeping the water requirement of rice in mind the Apa Tanis pay maximum attention towards this aspect. Development of better water management and irrigation system have helped the Apa Tanis to grow better crop of rice in their fields. Apart from water rice requires approximately 70°F temperature during the growing season for nearly 4-6 months. Clay soil is also necessary so that the surface water should not runaway by seepage. The Apa Tani's Valley fulfills all these ecological conditions. Located at the altitude of 1600 metres Apa Tanis Valley receives nearly 200″of rainfall per annum and that also coming in 4-5 months (last of May to October) by the South East Moonsoon. Heavy rainfall helps the Apa Tanis to submerge their rice fields with 5″-6″ water. However, heavy rainfall creates some problem like flooding.

Types of Rice Fields

The importance of rice field (Ajee) is extremely high in Apa Tanis society because from here comes the bulk supply of food. Rice field is also a symbol of social prestige. The richer and highly placed Apa Tanis have sufficient numbers of rice fields under their command. Land is the source of wealth and all other and less permanent possessions are mainly valued as a means of acquiring land (Furer-Haimendorf, 1946). Each and every Apa Tanis desires that he should acquire more rice fields. Here he differs from the neighbouring tribe Dafla (Nishis) and Hill Miris because in other societies land is not a personal property as it is in the Apa Tanis society. So it is necessary that while analysing the system of rice cultivation among the Apa Tanis little attention must be paid about rice fields (Ajee).

The geographical layout of the valley has laid a strong impact on the pattern/type of rice fields of the Apa Tanis. Furer-Haimendorf (1962) has classified the Apa Tanis rice field into two categories : In first category falls those rice fields which are kept permanently under water for most part of the year, whereas in the second category those rice fields fall which are occasionally flooded with water. However, my observation is this that the rice fields of the Apa Tanis may be classified into two broad categories on the basis of fertility and other agro-conditions and these two broad classifications may again be subdivided. These classifications are as :

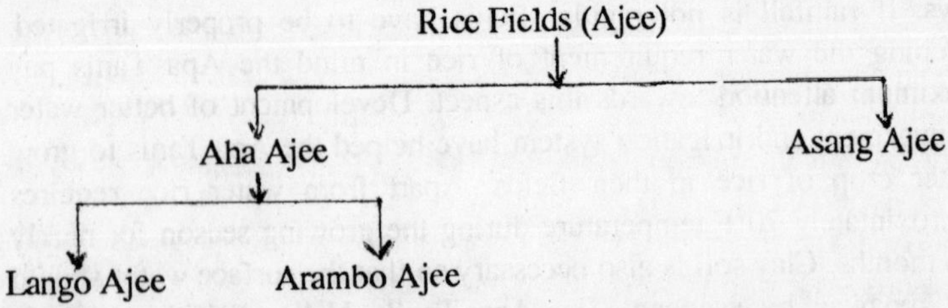

(1) *Aha Ajee* : In this category there are two sub-category which are as follows:

(a) *Lango Ajee* : The rice fields which are located very near to the village are termed as Lango Ajee. These rice fields receive the outlet water (discharge) from the village. The discharge water of the settlements carries lot of waste which later on are discomposed and increases the fertility of rice field. This type of rice fields are not more and there is extreme competition among the Apa Tanis to acquire more and more Lango Ajee. In Lango Ajee Pyaping and Emo variety of rice is grown.

(b) *Arambo Ajee* : If we proceed from the settlement the valley becomes more and more wider and the rice fields located here are called Arambo Ajee. The size of the Arambo Ajee rice fields become larger than the Lango Ajee. These rice fields are often flooded and each and every year new soil is added. However, it has been reported that its fertility is lesser than the Lango Ajee. Apa Tanis use to throw the household waste like rice husk (Piina) and ash (Mubu). It is common scene in the morning of an Apa Tanis Village where women can be seen carrying "Piina" and "Mubu" to the rice fields after the rice is harvested. These waste material add manure to the rice field. It is important to note that Apa Tanis do not use fertiliser of any sort. It is important to have this type of rice field because bulk of the paddy comes from these fields.

Kombu Ajee : This type of rice fields are located to the upper areas of the valley in between the hillocks (Putu) and hills (Pango). Here the rice fields become smaller and narrower because the valley itself becomes narrower. The rice fields situated in these areas are less fertile because soil contains boulder and sand. Due to upper elevation run off rate of water is also high which affects the rice production.

Method of Rice Cultivation

We have noted above the type of rice fields the Apa Tanis have to their command. It was important for the understanding of rice cultivation method because the type of rice grown in different types of rice fields differs. Attention in preparation of rice fields also differs according to their type and location. Preparation of rice fields takes major part of the Apa Tanis agricultural labour. Throughout the Winter and Spring seasons Apa Tanis repair the embankments and bunds of the rice fields. Apa Tanis women, children and also men do this job. Wherever it is necessary small boulders are placed with the help of bamboo poles. The second job in this process is the repair of water channels so that each and every field gets sufficient water. For all types of earthworks and also for carrying earth from one portion of the rice field to the other a flat wooden tray (Simpaya) is used. Spade (Dipe) is also used for various types of earthwork. In fact, spade (Dipe), wooden tray (Simpaya) are the most important agricultural implements of the Apa Tanis.

Apa Tanis have conserved their traditional mode of tilling the land. The main focus of the method is absolutely no use of animal energy. Even at this stage of development plough is not used though Apa Tanis are aware of it. There are different methods of tilling the land and thus vary according to the nature of rice field. The rice fields which are submerged under water throughout the year are not tilled. Roots of previous year are left and during the current session seedlings are planted in between the old roots/shoots. This is a character of the marshy land. In these type of rice fields late variety of growing rice is grown. The second category are those rice fields which remain wet through out the year but before the agricultural session starts some time in late winter water is drained and the field is allowed to be dried up. Then with the help of the spade, field is thoroughly dugged up and water is released. Before sowing, the rice field is again puddled by young men and women till the soil is fully changed to a smooth thick paste. Afterwards seedlings are planted. The last category of rice fields which are located on the uplands are dugged with spade, submerged under water and then the seedlings are transplanted. In the above two types of rice fields early growing variety rice is grown.

Apa Tanis grow rice through transplanting and this requires that rice must be grown first in the nursery bed (Miding). Nursery beds are located very near to the settlement and care is taken throughout the year to maintain the nursery plots. The bunds are regularly repaired and water is given properly. Kitchen ash are regularly spread in the nursery bed so that the fertility is maintained. For each and every variety of rice separate nursery beds are prepared. By the end of March the rice plants in the nursery bed grow upto 6"–8" and by the first week of April seedlings are ready to be transplanted in the rice fields. Generally 60-75 days old seedlings are transplanted.

Type of Rice Grown : *(Rice Emo)*

1. Empu Aare (Early variety of white rice) :
   - (a) Pyate (Sowing-M/L April) (Harvest-M/October),
   - (b) Pyaping (Sowing — M/L April) (Harvest — M/October),
2. Empu Hate (Late variety of white rice) :
   - (a) Empu (Sowing — L/April) (Harvest — L/October),
   - (b) Rade (Sowing — L/April) (Harvest — L/October),
3. Elang (Red variety of rice) :
   - (a) Pyare Sowing — M/L April).

(M = Middle, L = Late)

Apa Tanis grow basically 3 broad types of rice such as Empu Aare an early growing white rice, Empu Hate late growing white rice and Elang a red growing rice. This is again further subdivided. We will try to analyse this broad classification of rice in little detail. Empu Aare is a combined name of a group of rice which are known as early growing rice in the Apa Tanis agricultural system. In Empu Area there are mainly 2 varieties of rice such as Pyate and Pyaping. These varieties of rice are grown in the rice fields which are not submerged under the water throughout the year. The rice fields are dug properly with the help of spade and then water is released. Then young men puddle it and the mud is turned into paste. Transplanting of rice seedlings is done by middle of April. Transplanting is the work of women, young children and occasionally of men. Each and every seedling is planted at the interval of 8" approximately. The first variety to be planted is Pyate which is followed by Pyaping. By the middle of April transplanting of early growing rice is completed.

After the early growing rice Empu Aare, Empu Hate is transplanted. Empu Hate is a late growing rice in the Apa Tanis agricultural system. In this category there are two varieties of rice Empu and Rade. The late ripening varieties of rice are the main crop of the Apa Tanis. These varieties of rice are grown on the permanent wet rice fields which are marshy in nature. Here the fields are not dug with the help of spade. Men puddled the wet field till the soil is fully churmed and then it is levelled with a help of stick.

Apart from these two varieties of white rice Apa Tanis also grow a red variety of rice which is known as Elang or Pyare. Elang or Pyare variety of rice is grown mainly in Kombu Ajee. The method is same for this variety of rice also. However, most of the rice bear of the Apa Tanis is prepared with Elang or Pyare variety of rice. Weeding of rice fields are done with utmost care. Generally 4 times weeding are done after transplanting but this differs from rice field to rice field.

The first rice to be harvested is Pyare which is a red variety rice and harvest starts in early August. This rice is not cut but the grains are stripped from the ears by hand (Furer-Haimendorf, 1962). After Pyare the next rice to be harvested is Pyate which is a white variety rice. The method is same as it is done in the case of Pyare. But the main harvest starts in October. Men, women and children labour hard for at least two weeks to complete the job. The rice grain is cut with the help of seeckle and the stem is left in the field to get dried up and later on it is composed to become manure. After the harvest the crop is carried into basket and are brought to the thrashing ground where it is thrashed on a wooden plank. After that the grain is carried to the granary house (Nasu) for storing. The straw is left near the thrashing ground or stacked near the house/dry field/kitchen garden or thrown in the rice fields. Previously the straw was used for thatching the roof of the house but now roof is rarely thatched with straw. In the granary house rice grains are separated for seeds and for eating purpose.

Water Management

Apa Tanis plateau located in between 27° to 27° 38′ north latitude and 93° 45′ to 93° 55′ longitude receives heavy rainfall (200″) per annum from South East Monsoon which occurs from June to October. Due to heavy rainfall discharge of water through various streams and rivulets is very high. If the Apa Tanis want to grow good crop of rice ecologically it becomes important for them to develop good water management system. Through generations Apa Tanis have learned this art and today their water management system has no parallel in whole of North East. The basic factor in Apa Tanis water management is the use of bamboo for construction of small dams and also for water distribution system.

Every small and big streams which rise from the wooded high hills surrounding the Apa Tanis plateau are trapped and water is taken to various rice fields with the help of contour. Some time the stream is dammed at the upper head with the help of bamboo sticks so that the water level is elevated and can be channelised to the rice fields. This system is known as "Mugo Sigang Siling". After Winter session the irrigation system is thoroughly repaired so that it can last for 5-6 months when there is heavy flow in the streams due to heavy rain. Care is also taken so that no rice field gets flooded and damaged. In case when water has to be released from the paddy field a bamboo hole (Hubroo) is used. The whole water management of the Apa Tanis revolves around three important factors : Mutual distribution of water, Prevention of flood and soil erosion.

Conclusion

Apa Tanis rice cultivation has achieved a distinctiveness due to the various methods the tribe has developed keeping in view the typical natural features of the plateau on which it dwells. There is no doubt that it could not have assumed it in present form in any other part of Subansiri Region (Furer-Haimendorf, 1962). The intensification in growing rice has helped the Apa Tanis to maintain such a large population in such a small area. Such a development is mainly due the selfcontainness of the society both politically and ideologically and also due to sustained peace. Due to this character of the society problem regarding sharing of water has not occurred. However no water rights are sold but water reaches to each and every field in time and in appropriate quantity. Thus the Apa Tanis society can be a model to various other societies of Arunachal Pradesh.

REFERENCES

Adams, Inez., 1948 : Rice Cultivation in Asia, *American Anthropologist*, Vol. L. pp. 256-82.

Bender, S. J. and Wright G. A., 1988 : High Altitude Occupation, Cultural Process and High Plains Pre-history : Retrospect and Prospect, *American Anthropologist*, Vol. 90, No. 3, pp. 619-39.

Bower, U. G., 1953 : *The Hidden Land*, John Murry, London.

Chowdhury., J. N., *The Hill Miris of Subansiri*, North-East Frontier Agency Administration, Shillong.

———, 1973 : *Arunachal Panorama : A study in Profile*, Arunachal Pradesh Administration, Shillong.

Clyde, W., 1951 : The Rice Problem in Eastern Asia, *Geographical Magazine*, Vol. XXIV, pp. 11-21.

Elwin, 1957 : *A. Philosophy for NEFA*, North East Frontier Agency Administration, Shillong.

Furer-Haimendorf — Christoph Von., 1946 : Agriculture and Land Tenure among the Apa Tanis, *Man in India*, Vol. XXVI, pp. 20-49.

———, 1946 : Cultural Types in Assam Himalayas, *Indian Geographical Journal*, Vol. XXI, pp. 1-9.

———, 1955 : *Himalayan Barbary*, John Murry, London.

———, 1962 : *The Apa Tanis and Their Neighbours, A Primitive Civilisation of the Eastern Himalayas*, Rout Ledge & Kegan Paul, London.

———, 1980 : *A Himalayan Tribe from Cattle to Cash*, Vikas Publishing House Pvt. Ltd., Delhi.

Grist, D. N., 1953 : *Rice*, London.

Jen, Mei-Hago., 1948 : Agricultural Landscape of South-Eastern Asia, *Economic Geography*, pp. 157-69.

Izzard, R., 1951 : *The Hunt for the Bru*, London.

Jones, C. F., and Darkendwald, G.G., 1954 : *Economic Geography*, The Macmillan Company, New York.

Kuriyan, G., 1945 : Rice in India, *Indian Geographical Journal*, Vol. XX. pp. 28-36.

Sinha, S., 1985 : Social and Ecological Context of Rice Cultivation in India, *East Asian Cultural Studies*, Vol. XXIV, No. 1-4, pp. 83-92.

Singh, A. K., 1985 : Ecological Factors and Traditional Agriculture among the Chettis of Nilgiri Wynad, in Chakerborty, S. B. and Others (Eds.), Agrarian Situation in India, Vol. II, *Anthropological Survey of India*, Calcutta, pp. 109-22.

Sukla, B. R., 1959 : *The Daflas of the Subansiri Region*, North East Frontier Agency, Shillong.

Tado, W., 1985 : Origin and Dispersal of Rice in Asia, *East Asian Cultural Studies*, Vol. XXIV, No. 1-4., pp. 33-40.

Tuan, D. T., 1985 : The Types of Rice Cultivation and its Related Civilisation in Vietnam, *East Asian Cultural Studies*, Vol. XXIV, No. 1-4, pp. 41-56.

[illegible] and [illegible] 1959 [illegible] High [illegible] Kathmandu [illegible] Vol. 60 No. [illegible] pp. 518 [illegible]

[illegible] G. 1952 [illegible] Mary [illegible]

[illegible] 1962 [illegible] Frontier Agency Administration, Shillong.

[illegible] Frontier Administration, Shillong.

[illegible] 1957 [illegible] Geographical Magazine, Vol. XXIX, pp. 11-21.

[illegible] 1957 [illegible] North East Frontier Agency Administration, Shillong.

[illegible] Agriculture and Land [illegible] among the Apa Tanis, [illegible] Vol. XVI, pp. 20-47.

[illegible]

— 1975 [illegible] Asia Publishing House [illegible]

— 1980 [illegible] Frontier [illegible] Publishing House [illegible]

[illegible]

[illegible] Agricultural [illegible] Eastern Asia, Economic Geography [illegible]

[illegible] The [illegible]

[illegible] and [illegible] Economic Geography, The Macmillan [illegible], New York.

[illegible] The [illegible] History of [illegible]

[illegible] Vol. XIV No. [illegible]

Smith, [illegible] 1962 [illegible] in [illegible] Asiatic Society of Bengal [illegible] Calcutta, pp. [illegible]

[illegible] B. [illegible] The [illegible] North East Frontier Agency, Shillong.

[illegible] East [illegible] Vol. [illegible] pp. 33-39.

[illegible] Vol. XLIV, No. 1-4, pp. 1-15.

16

RURAL FLOOD MITIGATION PROJECT ASSESSMENT

G. T. McDONALD

PROBLEMS OF planning an appropriate combination of structural and non-structural measures to mitigate flood damages in rural areas have received a great deal of consideration but difficulty still exists in providing suitable technical data to allow adequate evaluation of alternatives and to indicate the extent of hydrological change in the planning region. Agricultural geographers can help in this work. Of course other more intractable problems exist such as scaling the long-term economic significance of any benefits, accounting for secondary effects and forecasting the impact of changes on the long-term equilibrium of flood-plain habitats. The research reported here is concerned with a geographical data base and analysis suitable to both assessment of the proposed mitigation schemes and their design. It can also be used to shed some light on the less conventional decision-making problems just mentioned. Most of the research described here can be found in a number of reports that assessed the hydrological, economic and environmental impacts of a project on the river Macleay flood-plain, N.S.W., Australia (McDonald, 1967, 1968; N.S.W. Department of Public Works, 1978).

The river Macleay flood-plain is typical of many flood-plains throughout the world. The river reaches tidal influence just above the town of Kempsey, from which point it meanders through a system of high natural levees to the sea some 45 km. downstream. The flood-plain grades back from these levees and those of the three major distributary streams, Belmore River, Clybucca Creek and Kinchela Creek to what used to be large semi-permanent swamps. These depressions along with a number of other minor depressions formed by earlier meander movements are isolated below the height of the stream by the levees and became naturally land-locked. The width of the flood-plain varies, reaching its widest section of 20 km. midway between Kempsey and the sea. At the estuary a large area of mixed swamp land, tidal marsh and dune has been created by the migration of the river mouth.

Climatically the area is suited to a range of pastoral and agricultural activities that are concentrated on the better drained areas of the flood-plain where soils and topography are excellent (Walker, 1963). Man has elected to settle and utilise an area with some first-class agricultural qualities, but an area subject to periodic flooding: minor flooding one year in three and major flooding one year in five. Flood control is not feasible in this area and a mitigation of damages was the only practical proposition.

Briefly, the works proposed were:

(i) By a system of levee-bank improvements along the river and distributaries and the installation of headworks across creeks and drains, minor floods could be completely contained. ($Q < 2000$ m^3/s).

(ii) Floods with peak discharges of between 2,000 and 3,000 m^3/s could be contained within the river system and several natural floodway storage areas. These floodways needed some improvement work but the areas so used were very low-lying swamp land serving this purpose under natural conditions.

(iii) By construction of drainage works and associated headgates, floodwater could be removed more quickly once it had inundated the flood-plain.

(iv) Erosion of the river bank was minimised by bank protection works. In some cases this was expected to improve the capacity of the channel.

(v) Four additional flood outlets to the sea were constructed to assist in the more rapid drainage of major floods. Two of these entered existing small creeks and two were entirely new cuttings through the contemporary dune system (Fig. 16.1).

It was in this area that the project assessment model described below was developed. It proved to be most useful and has been adapted to other areas.

The Project Assessment Model

The model has been described in the accompanying Fig. 16.2 and Table 16.1. Detail is not important since the results apply to the Macleay flood-plain. What is important is the set of components outlining the major research steps required in understanding the environment and man's adoption to it. It may not be feasible to carry out all these steps in every case and perhaps more approximate measures can be used. Basic hydrology is required for some of the steps and a multi-disciplinary team might be required.

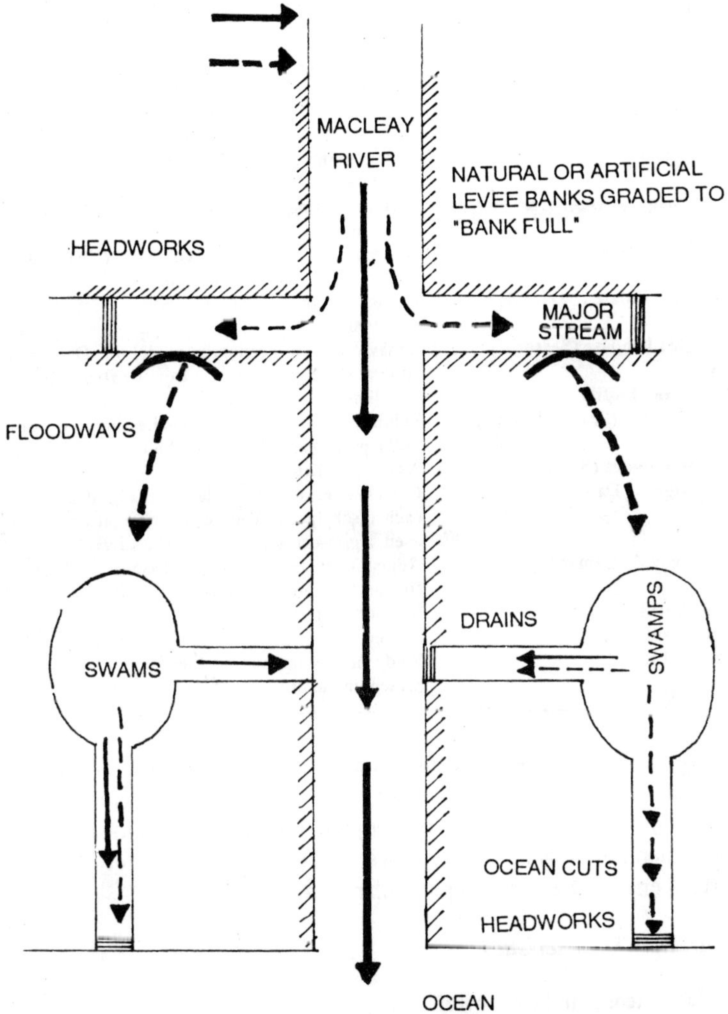

FIG. 16.1: Basic design characteristics of the flood mitigation system.

Assessment Applications

Four types of analysis will be undertaken :

(i) Assessing the distribution and extent of flooding characteristics;

(ii) Comparing the damage sustained before the installation of mitigation works with that existing now;

(iii) Measuring the changes brought about by parametric variation in the design capabilities of mitigation works;

(iv) Commenting on possible changes in other environmental variables and their effect on other land functions.

TABLE 16.1: Components of the assessment model

Component	Research Technique
Flood-plain Unit	Topographic survey. Scale approx. 1:5,000 — 1:10,000.
River Stage — Discharge	Establish rating curve for the river at the gauging point: height/flow.
Flood Recession	Drainage channel flow volumes (observed or predicated) and unit hydrographs.
Existing Landuse (Seasonal)	Landuse survey. Scale approx. 1:5,000 — 1:10,000.
Projected Landuse (Seasonal)	Behavioural or economic model or Descriptive approach using information from local leaders.
River Bank Profile	Levelling.
Crop Damage (Seasonal by Type)	Published research. Questionnaire survey of farms in the project area. Enterprise budgets.
Pasture Damage (Seasonal by Type)	As for crop damage.
Discharge — Damage	Calculate area flooded landuse and damage for each discharge. If computer techniques are being used, a grid-cell scheme would be valuable.
Frequency — Damage	Adjust damage figures by seasonal and annual frequency to provide expected annual damages.
Soil Classes	Soil survey 1:10,000.
Water Table	Survey.
Aquatic Life	Study of biology and ecology of acquatic species present in the project area.

These analyses will be undertaken for the particular study area summarised in Fig. 16.2 as an illustration of the value of the methodology to the planning of mitigation works but also to show how such quantitative analysis helps describe the relationship between man and his environment in this particular area.

Flooding Characteristics

(a) Extent of turbulent overflow :
The change is readily seen in Fig. 16.2(i). The length and depth of the flow has been dramatically altered by the reinforcement of levees in low-lying sections. The extent of flow of greater than two-feet depth in a major flood has been reduced to zero from a total length of almost two miles in six different locations.

(b) Frequency of inundation:
The maximum frequency has been reduced from annual to less than once in two years for the lowest-lying areas. Most of the area is still

flooded with a probability of once in five years; the degree of improvement proportional to the level of the original risk.

(c) Rate of drainage
The entire area is now drained in seven days [(Fig. 16.2(f)]. Some areas had a marginal improvement of a few hours but the outstanding feature is the more rapid drainage of land below R.L. 106. This affects some 286 ha of swamp pasture and is more of a reclaimation than a flood hazard reduction benefit to such areas.

Flood Damage

(a) Existing flood damage :
As shown in Fig. 16.2(l) the damage caused by relatively small floods is high. This is a result of the stage-discharge and topographic characteristics of the particular flood-plain and the relatively intensive use of flood-prone areas. Damage to crops is highly seasonal and related to flood levels but taken together with pasture damageability and long recession times the increase in losses for larger floods is not great. (Loss of life, stock and buildings has *not* been included).

When frequency adjusted, the importance of the smaller floods is more pronounced [Fig. 16.2(m)].

It is possible with this technique, which is a predictive device only as good as the basic damage relationships shown in Figs. 16.2(j) and 16.2(k) to disaggregate damage by crop type by land area and by season. The estimates will have a significant variance due to the combined effect of excluded factors such as temperature, turbidity, prior events, etc.

(b) Post-project flood damage :
A given set of levees, drains and diversions was incorporated into Figs. 16.2(e), 16.2(d) and 16.2(f). The discharge damage and frequency damage curves, Figs. 16.2(l) and 16.2(m), show the effects of that combination. It is clear that the concepts were highly beneficial in reducing the frequency of flooding, Fig. 16.2(e) preventing it crossing the intensively settled levees, Fig. 16.2(i), and removing it within a design period of seven days, Fig. 16.2(f). Damage for smaller floods was thereby prevented and for medium-size floods substantially reduced. For the largest flows damage savings are not great.

Parametric Variations

Given the systematic basis of this project assessment model it is possible to alter the size of single elements, for example, a given levee, drain or diversion and determine the hydrological and economic effects. This is extremely useful for planning purposes and the model has been the basis of attempts at optimising the project.

Effect on other Land Functions

Apart from landuse an area of land has further social usefulness in providing sites for "Nature's Activities", for biophysical processes such as water purification, wildlife breeding and gene-pool maintenance. They are important for man's continued existence and in flood-plains impacts on these functions must also be mentioned (Austin and Cocks, 1978).

Given the nature of proposed engineering works, the changes in other land functions will be caused by hydrological changes resulting from :

(a) Reduced incidence of inundation. This can be quantified for any area (grid cell).
(b) More rapid drainage and more fresh water in river following all floods.
(c) Change in the level of and flow in the water table.
(d) Reduction in the frequency of seasonal fresh swamp conditions.
(e) No intrusion of salt or brackish water up distributary channels.

All these changes can be quantified in terms of area (grid cell), levels, landuse affected, time and probability.

These changes are in themselves inconsequential but the effects they have on other land functions are relevant. Some effects indicated from the model and other research are :

(a) A reduction of 2,000 hectares of freshwater wetlands important for birds (Goodrich, 1970). Furthermore, it seems that many aquatic birds require oscillating water levels not just regulated areas of fresh water indicating they are adapted to periodic flooding. Minor changes in the drain design can be evaluated both with respect to the possible retention of areas of wetlands and the economic consequences to other landuses.
(b) A reduction of 2,100 hectares of saltwater plant communities including mangroves, salt marsh and sea grass as a result of particular levee and river training works. These areas and associated water bodies are important as fish-breeding areas for an estimated 70 per cent of the areas commercial fish species.
(c) The alteration of river flow during flood times, especially the longer period of fresh-water flow in improved drainage works, may be detrimental to oyster populations known to have a very limited tolerance for longer periods in fresh water.
(d) Lowering of the water table resulting from levee and drainage works can have complex effects on soils and water tables. O'Brien has recommended maximum limits on the depth of drainage to RL 101 so that the estuarine clays are not subject to drying. "Wetting and drying

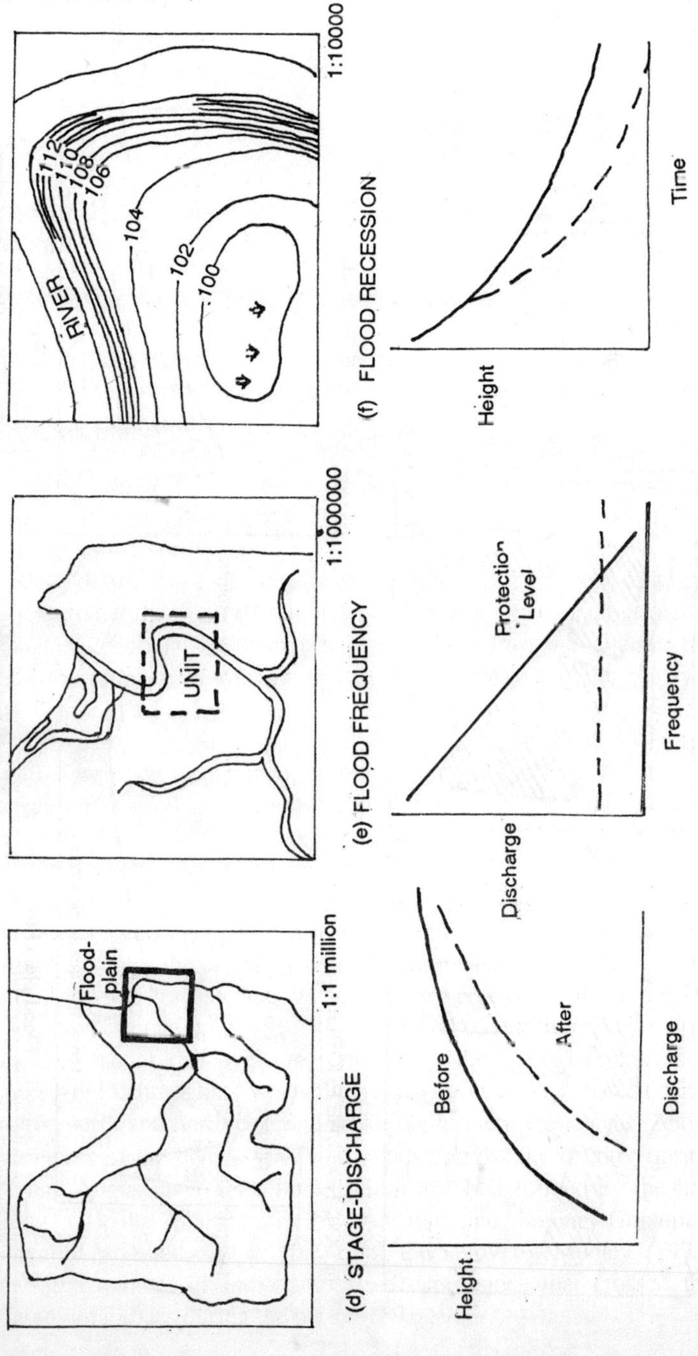

FIG. 16.2: Rural flood mitigation project assessment model.

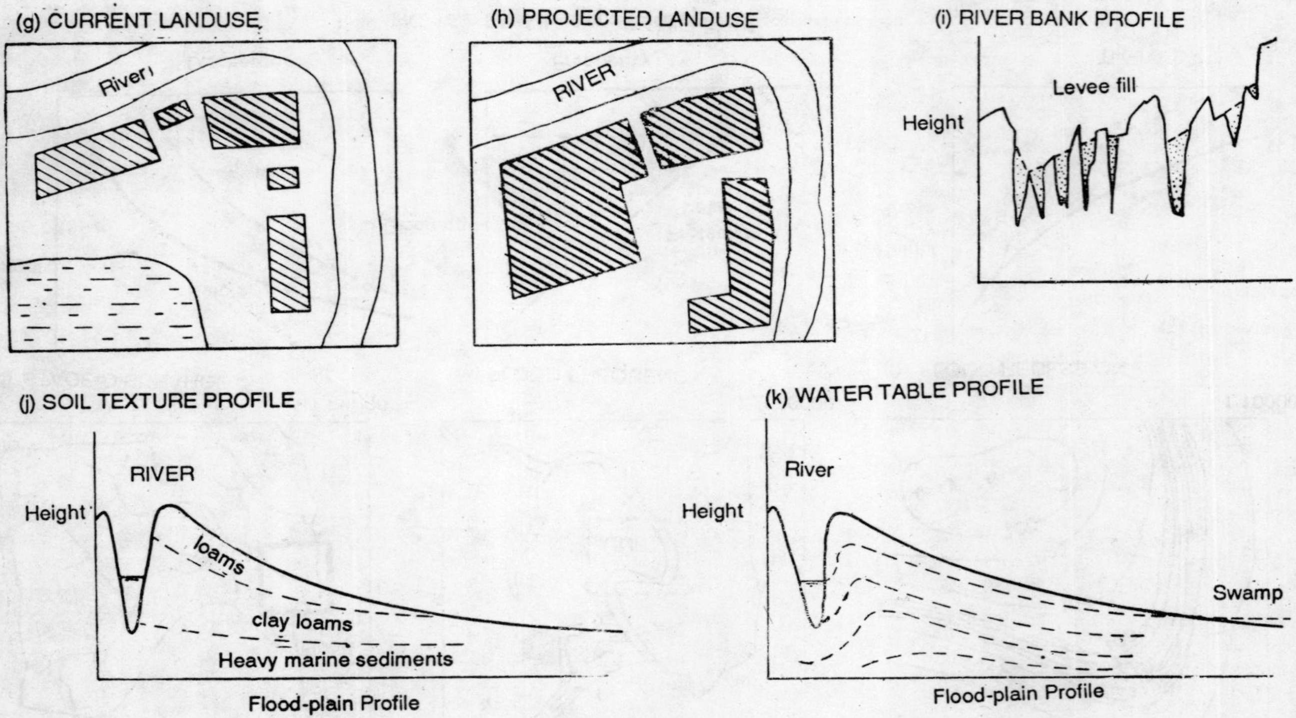

FIG. 16.2: *(Contd.).*

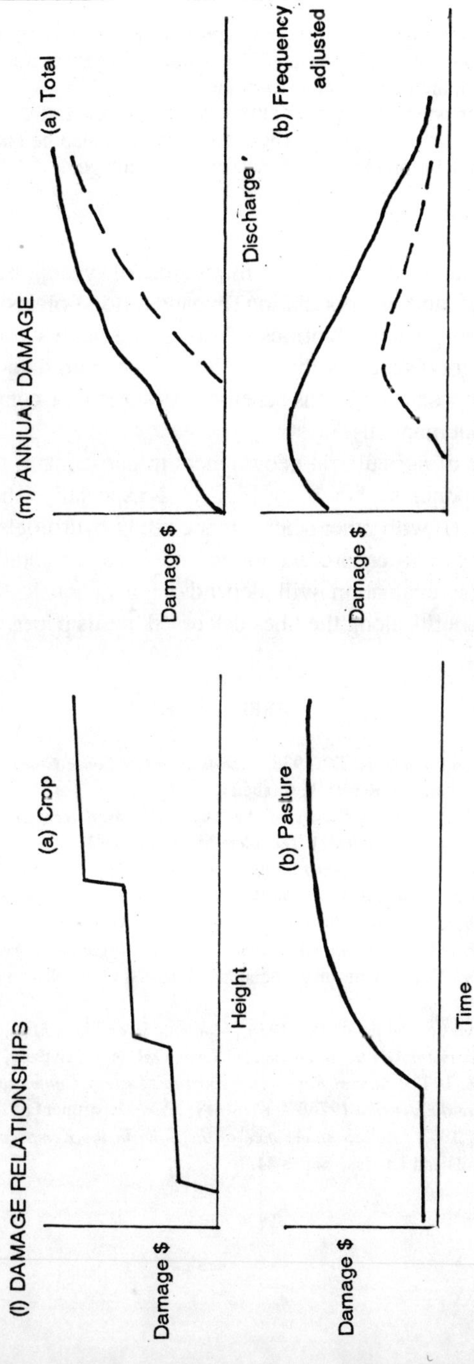

FIG. 16.2: *(Contd.)*.

cycles in those soils produce rapid changes in acidity in the soil and in the soil water" (O'Brien, 1977). These changes are not easy to quantify nor are their effects easy to evaluate.

Lower water tables also allow salt water penetration into levels used for irrigation purposes. These changes must also be taken into account when determining the optimal level of drainage.

Conclusion

Man has managed rivers for thousands of years, but not always successfully, and as population expands flood-plains will need to produce even greater quantities of flood. Research workers can assist in creating good schemes through field survey and data analysis along the lines proposed here. This scheme is designed for computers but can be carried out manually.

The role of agricultural geographers in particular is to work on the model components shown in Fig. 16.2 especially 16.2(c), 16.2(g), 16.2(h), 16.2(l) with other scientists including hydrologists, economists and pedologists to ensure economic and environmentally harmonious projects. The evaluation will depend on a quantification of project costs and benefits along the lines discussed in this paper.

REFERENCES

Austin, M. P. and Cocks K. D., 1978 : *Landuse on the South Coast of N.S.W.*, CSIRO Division of Landuse Research, Canberra.

Goodrick, G. N., 1970 : *A Survey of Wetlands of Coastal New South Wales,* CSIRO Division of Wildlife Research Technical Memorandum No. 5.

McDonald, G. T., 1967 : *A Report on the Hydrological Implications of Flood Mitigation Works on the Floodplain of the Macleay River Below Kempsey,* Department of Geography, University of New England Armidale, Australia.

——— 1968 : *An Economic Appraisal of Flood Mitigation Works on the Macleay Floodplain,* Department of Geography, University of New England Armidale, Australia.

NSW Department of Public Works, 1978 : *Macleay River Flood Mitigation Environment Impact Statement Additional Drainage: Seven Oaks Area,* Sydney, Australia.

O'Brien, A. D., 1977 : *Annual Research Report to Macleay Committee for Agricultural Research and Extension 1976/77,* Kempsey, NSW Department of Agriculture.

Walker, P. H., 1963 : *A Reconnaissance of the Soils in the Kempsey District,* N.S.W., CSIRO Soils and Landuse Series 44.

17

FERTILITY AND PRODUCTIVITY OF SOIL

N. K. DE

SOIL IS the basis, the foundation of farming. Without it we get nothing. with poor soil, poor farming, poor living, whereas with good soil, good farming and good living. An understanding of good farming begins with an understanding of the soil. Through understanding, poor soils can be made better, and good ones better still. Just as careless and unwise use might soon lower the value of even a very good soil and, perhaps, reduce its production for a very long time. Soils differ in many ways and what is a poor soil for one crop may be very suitable for another. Thus, understanding soils may involve getting to know the best use for many different types of soils, each used and managed according to its individual characteristics. There are different aspects of soil, and its use and fertility and productivity is only one aspect of it which has been discussed in this paper.

Soil Fertility

The concept of soil fertility is very complex. There is no absolute scale of fertility because of the consideration, the use to which land/soil is put. The term fertility is used in different senses in different contexts.

Soil fertility is the quality that enables the soil to provide the proper compounds, in the proper amounts and in the proper balance, for the growth of specified plants, when other factors, such as light, temperature, moisture and the physical conditions of the soil are favourable. It is thus the capability of soil to produce a plant yield under defined conditions.

From the standpoint of agriculture soil fertility is best understood by : (a) the nutrient requirements of plants, (b) the supply of nutrients in the soil, (c) the ways in which nutrients are lost from the soil, and

(d) methods by which soil fertility is restored or maintained.

Soil Productivity

Soil productivity is the capability of a soil for producing a specified plant or sequence of plants under a physically defined set of management practices. It is measured in terms of inputs of production factors in relation to outputs or harvests. Thus soil productivity is not itself an inherent quality of the soil. All the chemical, physical and biological properties of a soil, together with the associated climate, determine its response to management — response to inputs of labour and materials.

Two soils may each give low yields with a set of simple practices — mere tillage, seeding and harvesting. The soil productivity of both may be low, yet one may respond far more to inputs of fertiliser and time, and to the use of rotations with deeply rooted legumes than the other. One soil has, with this class of management, high productivity while the other remains low.

Thus the term "natural soil productivity" cannot be defined. The expression is in fact meaningless. It is only in a cultural setting that soil productivity exists or can be measured. Whenever people use loosely the term "natural soil productivity", they have in their mind already an unconscious assumption of management that is common to the area. Certain tropical soils, for example, in one cultural setting are said to be unproductive, while in another setting they are highly productive.

Thus soil productivity and yield predictions are achieved in two principal ways : (a) through judgements based upon evidence afforded by actual yield data from sample areas of the soil-mapping units, and (b) through judgements based on comparisons of the characteristics of soils and basic knowledge of plant requirements.

Factors Controlling Soil Fertility

Soil fertility is influenced by the countless individual factors which are inter-linked in such a way that they cannot be rigidly separated. However, for our convenience they may be grouped as follows :

1. Soil depth.
2. Soil texture, soil structure, soil pore space, soil temperature, soil aeration, soil compaction and tillage.

3. Soil reaction.
4. Nutrient content.
5. Humus content.

Soil Depth

The thickness of soil layers readily occupied by plant roots determine the effective soil depth. The soil depth varies from place to place (more than 90 cm to less than 25 cm). In a deep soil, plants withstand drought better because the roots are spread through a larger volume of soil and also plants can absorb the minerals stored in the sub-soil if their roots can reach them.

Soil Texture, Soil Structure, Soil Pore Space, Soil Temperature, Soil Aeration, Soil Compaction and Tillage

Soil Texture

The relative size distribution of soil separates or particle size groups, is referred to as Texture. The common textural classes are sandy loam, loam, silty loam, sandy clay loam, clay loam, silty clay loam, silty clay, etc.

Soil Structure

The primary soil particles are bounded together in varying degrees into larger units of semi-permanent clusters or groups of soil particles. The shape, size, arrangements and stability of these aggregates collectively determine the structure of the soil. The common structural classes are — granular, platy, blocky, prismatic, etc.

By affecting porosity the texture and structure of the soil change the physical and chemical environment in which the plant roots grow. The physical properties of the soils (texture, structure, etc.) influence crop growth through their effect on soil moisture, soil air, soil temperature and impedance to root development and shoot emergence. If the physical conditions of the soil are not favourable to water or plant roots to penetrate and to the germinating seeds to break through the soil crust, the crop yields will necessarily be reduced, even though the soil be very fertile.

Soil Pore Space

A random distribution of the primary and secondary soil particles results in a spectrum of pore size distribution congenial or uncongenial to good plant growth. The volume, dimensions and the shape of soil pores are the decisive factors governing the behaviour of soil moisture and soil aeration.

A well-aggregated soil possesses a large-pore-space, allows the penetration of rainfall and gaseous exchange to the subsoil region. Such soils offer less mechanical impedance to root development and shoot emergence and do not form crusts easily.

Soil is a three-phase system, containing solids, liquids and gases. The more the water, the less the room for air, and vice-versa. A balance between liquid and gaseous phases for the healthy growth of roots requires to be carefully corrected by cultural practices for optimum crop yields.

Studies of porosity and pore-space distribution help in planning crop rotation and cultural practices.

Soil Temperature

Nutrient absorption in soil will be affected by temperature that influence metabolic activity. Many investigators have indicated that the absorption of solutes is reduced at low temperature. High temperature results in salt accumulation in soil. Low temperatures do not appear to retard seriously the absorption of N, but they do affect the capacity of the roots to reduce the absorbed nitrate and convert it into organic nitrogenous forms.

Soil temperature, which does not inhibit root permeability and plant growth, may either increase or decrease *P* availability to plants. In organic soils, *P* availability may be increased by low temperature. In many soils an increase in temperature may increase the rate of mineralisation of the organic matter, thus making more *P* available to plant. Also a temperature change during the year may cause the shift in the microbial population which may affect the mineralisation of *P* availability.

Soil Moisture

The soil moisture content has an intimate relationship with the nutrient availability. The *N, P* and *K* contents of the plants are appreciably higher under moisture stress, but the reverse relationship is observed for calcium.

Soil Aeration

The presence of an insufficient quantity of oxygen or high concentration of CO_2 result in restricted nutrient uptake by the roots of a large number of plant species. The accumulation of ions in the root tissue is strangely influenced by the concentration of oxygen.

Soil Compaction and Tillage

Compacted soils restrict root penetration and aeration and, thereby, exert a direct effect on the ability of the plant roots to extract both nutrients and moisture from soil. Furthermore as compaction reduces infiltration, it decreases the utilisation of precipitation. As a result crops may experience both a moisture stress and a shortage of plant nutrients. Mechanical measures can be employed to break through compacted soil layers or horizons.

Tillage promotes the oxidation of the incorporated organic residues and, thus, facilitate mineralisation and the release of nitrogen. Thus under the minimum tillage system additional quantities of N will be required.

Increased compaction has been observed to decrease *P* absorption by limiting soil aeration. As pore sizes decrease owing to compaction the roots are not able to penetrate the soil easily.

Soil Reaction

Soil reaction or the soil pH is the acidity or alkalinity of the soil. Soil pH influences the rate of plant nutrient release by weathering, the solubility of all materials in the soil, and the amounts of nutrients stored on the cation exchange sites. The pH is, therefore, a very good guide for predicting which plant nutrients are most likely to be deficient.

Most nitrogen and sulphur mineral compounds (except calcium sulphate) are highly soluble at any pH and subject to rapid leaching. The rate of mineralisation of these elements from organic matter is fastest between pH 6 and 8.

A pH between 6.5 and 7.5 is usually best for phosphorus

availability. Potassium is soluble at any pH but is removed from solution by sorption. Iron availability generally increases as the pH is lowered. Manganese, copper and zinc may be leached out and become deficient at very low pH. Boron may be leached at very low pH. Molybdenum solubility is increased by iron and aluminium at low values.

A pH value near neutral is best for most plants. The availability of all the plant nutrients is at least reasonably satisfactory at pH values between 6.5 and 7.5.

Nutrient Contents

Depending upon the quantitative requirements, the essential elements (nutrients) are usually grouped under two categories : (a) Macro-nutrient elements, and (b) Micro-nutrient elements. Elements which are required by plants in concentrations exceeding one part per million, often ten times or more, are called "macro-nutrients". These include carbon, hydrogen, oxygen, nitrogen, phosphorus, potassium, calcium, magnesium and sulphur. Elements which are required by plants in concentrations less than 1 p.p.m. are put in the category of micro-nutrient elements, also referred to as "micro-nutrients" and include elements like manganese, copper, zinc, molybdenum, boron, chlorine, cobalt and iron.

The division of essential elements into macro and micro-nutrients is not always valid as the quantitative requirement of particular nutrient elements has been shown to differ in the different groups of plants and even indifferent varieties of the same species. Even the same plant may have widely different nutrient requirements at the different stages of its growth and differentiation.

The role of Macro-nutrient : The essential elements play important but different roles in plants. The functions of essential elements are : (i) as structural components of cell constituents and its metabolically active compounds, (ii) in the maintenance of cellular organisation, (iii) in energy transformation, and (iv) in enzyme action, carbon, oxygen and hydrogen — carbon, oxygen and hydrogen are obtained by plants from air and water. Carbon and hydrogen are absorbed in a combined form but oxygen is partly taken up in the molecular form. Apart from having a major structural role, carbon, hydrogen and oxygen play a key role in providing energy required for the growth and metabolism of plants.

Nitrogen

Nitrogen is a major structural constituent of the cell. Nitrogen plays an important role in plant metabolism. When the nitrogen supply is the limiting factor, both the role and the extent of protein synthesis are depressed. Excessive amounts of nitrogen are also harmful. High nitrogen produces succulence in plants and enhances their sensitivity to water and temperature stress. High nitrogen plants also become susceptible to lodging, pathogens and pests. Nitrogen supply in soil can take place through four ways : (i) Nitrogen fixation by symbiotic bacteria, (ii) azofication or N fixation by non-symbiotic bacteria, (iii) addition by rain or snow, and (iv) addition through manures and fertilisers.

The removal of nitrogen from the soil occurs : (i) by plants and micro-organisms, (ii) by loss of nitrate in leaching water, and (iii) by volatalisation.

Phosphorus

Like nitrogen, phosphorus plays an important role as a structural component of the cell constituent and metabolically active compounds. Phosphorus plays an important role in energy transformations and metabolic process of plants. It is also involved in the activation of a number of enzymes participating in the dark reactions in photosynthesis. Phosphorus deficiency is characterised by low organic P/inorganic *P* ratio in plants. The soil phosphorus exists in many primary and secondary compounds. The apatite group of primary minerals is the original source of about 95 per cent or more of the soil phosphorus.

Potassium

Potassium plays an important role in the maintenance of cellular organisation by regulating the permeability of cellular membranes and keeping the protoplasm in a proper degree of hydration by stabilising the emulsions of highly colloidal particles. Potassium deficiency is known to produce water imbalance in plants. Potassium salts have a great buffering actions and stabilise various enzyme systems. Apart from its organisational and electrochemical role, potassium plays a catalytic role. Potassium deficiency is known to bring about varied effects on nitrogen fractions in different species of plants. Potassium is known to increase the resistance of plant to the stress of moisture, to heat and to diseases caused by pathogenic fungi, nematodes and other micro-organisms. In cereals, potassium improves the proper development of mechanical tissues in the straw, making them less susceptible to lodging. In fruit trees, potassium improves the colour, flavour and size of the fruits.

More than 95 per cent of the total potassium in the soil is within the crystal lattices of silicate minerals. The micas and the feldspars constitute the major K-bearing minerals which, on weathering, slowly release K to the soil. The secondary clay minerals also constitute an important source of soil potassium.

Calcium

Calcium is involved in the maintenance of cellular organisation of plants. Calcium is suggested to play an important role in neutralising the acids, which may become injurious to plants. Calcium favours assimilation of nitrogen into organic constituents especially proteins and is an activator of a number of enzymes. Calcium is present in all soils in different forms. The calcareous soils have a higher amount of calcium than the acid soils.

Magnesium

Magnesium is needed by all green plants, as it is a constituent of chlorophyll and also takes part in phosphate transport. It is a very mobile element. In the case of its deficiency, it moves from the older leaves to the younger ones. Magnesium is also required for the activation of many enzymes. The magnesium content of soils depends upon the nature of the parent material, the degree of weathering, the texture of the soil, the intensity of cropping and management practices.

Sulphur

Sulphur is a constituent of the amino-acids, cystein and methio-nine and proteins. Being involved in the activation of a number of enzymes, participating in the dark reactions of photosynthesis. Sulphur is involved in carbohydrate metabolism of the plants.

In sulphur-deficient plants, sulphur is involved in the proper development of tissues, e.g. formation of hard and woody stem.

The earth crust contains about 0.06 per cent sulphur. The initial source of all sulphur is the sulphides contained in plutonic rocks. The other sources of sulphur in the soil are the atmosphere, fertilisers, organic manures, plant residues and irrigation water.

Iron

Iron is a constituent of porphyrin compounds. The haemoglobin of the leguminous root nodules contains iron as an essential constituent. In some micro-organisms, iron is a structural constituent of pigments. The metal is known to affect enzyme actively by being both a structural part and an activator of enzymes. Iron deficiency causes a decrease in the concentration of nuclic acids, sugats, organic acids, vitamins, etc. Iron is a constituent of biotite, hornblend and chloride. Indian soil contains 0.46 to 27.5 per cent of iron.

Manganese

Manganese may give a characteristic chocolate colour to the upper layers of soils, particularly in podzols. Manganese deficiency in soils may be overcome either by spraying the diseased plant or by applying manganese to the soil. Manganese has been found to exist in soils as water soluble exchangeable and high oxides of manganese which exist in equilibrium in soils. It occurs in large quantities (sometimes 1%) in some soils.

Copper

Copper is a constituent part of several enzymes participating in the cellular oxidation-reduction. Copper stress in plants has been shown to result in a decreased rate of photosynthesis. Copper stress has also been shown to affect other metabolic processes (the respiratory oxygen uptake), the activity of certain enzymes and the nitrogen metabolism. The normal agricultural soils of the world contain 1 to 100 ppm copper.

Zinc

Zinc plays an important role in both micro-organisms and higher plants. Zinc deficiency has been known to result in a decreased level of photosynthesis, in the accumulation of amide and α-amino nitrogen, reduced flower setting and delayed development of fruits. Zinc has been suggested to play a role in regulating the auxin concentration in plants. Average zinc content of mineral soil varies between 10 and 330 ppm.

Molybdenum

Molybdenum has been shown to be required for nitrogen-fixation. Molybdenum deficiency decreases concentration of sugars and the poor formation of middle lamella and also results in reduced levels of photosynthesis. The total molybdenum content of most of the soil varies from 0.2 to 5.0 ppm. Molybdenum is generally present at : (i) part of the crystal lattice of primary and secondary minerals, (ii) absorbed molybdate ion on clay or exchange complex, (iii) part of the organic matter, and (iv) water soluble compounds.

Boron

Boron occurs in soil in very small quantities, generally less than 20 parts per million, but it appears to be essential for most plants. A deficiency of boron is thought to be the cause of *brown heart* in turnips and of *heart rot* in sugar beet. Small amounts of boron are necessary in the formation of the nodules or leguminous plants and a deficiency interferes with the movement of carbohydrate that is essential to their formation.

Chlorine

Chlorine has been shown to be involved in the oxygen evolution in primary photosynthetic reactions. It has also been claimed to be involved in cyclic photosythesisation.

Chlorine is widely distributed in nature and large quantities of it are carried in land from the sea by wind and rain.

Cobalt

Cobalt has been shown to be essential for the growth and metabolism of diverse micro-organisms. In the absence of cobalt, legumes incubated with a proper strain of rhizobia fail to fix atmospheric nitrogen. It is a structural component of vitamin B_{12} — Cyanocobalamine synthesised by many bacteria. Cobalt may also play a catalytic role by being an activator of certain enzymes.

Sodium

In higher plants, sodium has so far been shown to be essential only for two halophytic species. Atriplex vasicaria and Hologeton glomeratus. Perhaps the best known role of the element is in the maintenance of osmotic relation of the cell. In sugar-beet, sodium exerts great beneficial effect — water relations are markedly affected by this element.

Humus Contents

Organic matter is an important component of the soil, which supplies the essential elements to plants after decomposition and improves the physical condition of the soil. The soil organic matter consists of a whole series of products which range from undecayed plants and animal tissues to fairly amorphous brown-to-black material bearing no trace of the anatomical structure of the material that is normally defined as the soil humus. In addition soil organic matter contains living and dead microbial cells, microbially synthesised compounds and derivatives of these materials produced as a result of microbial decay.

Humic and Non-humic Substances

Soil organic matter can be conveniently divided into two groups : (a) non-humic substances which include all those classes of compounds occurring in plants and micro-organisms that appear to have relatively definite characteristics. In this group can be included carbohydrates, proteins, fats, waxes, resins, pigments and low-molecular-weight compounds. Most of these could be easily attacked by soil micro-organisms and have a rapid turn-over in the soil, (b) humic substances which are materials produced in the soil and are either yellow or brown-to-black, acidic, polydisperse substances of high molecular weight. Based on solubilities the group is generally divided into three classes : (i) fulvic acid, (ii) humic acid, (iii) humin.

Role of Organic Matter in Soil Fertility

The importance of soil organic matter in relation to soil fertility and physical conditions is widely recognised. Soil formation is closely linked with the action of diverse forms of organic substances on the porent rock, and bio-geochemical process. In the formation of a fertile soil, organic substances play a direct role, as they are the sources of plant nutrients which are liberated in available forms during mineralisation. Humus can be considered to be a store-house of various nutrients essential to plant growth. During the slow microbial decomposition of the soil humus, there is a gradual release, with subsequent mineralisation of carbon, nitrogen, sulphur, phosphorus and other elements.

Besides being a source of nutrients for the plants, the presence of humus in soil exerts a pronounced influence on the physical, chemical and biological activities of the soil. In improving the physical conditions of the soil, there are two distinct facets of the problem. One is the built-up organic carbon status of the soil whereas in the other it is the process of decomposition in the soil. Humus improves the structure of the soil, its drainage and aeration, increases its water holding, buffer and exchange capacities, influences the solubility of soil minerals, and serves as a source of energy for the development of micro-organisms.

Factors Controlling Soil Productivity

Like soil fertility, soil productivity factors are also innumerable. As has already been stated soil productivity is not an inherent quality of the soil but all the physical, chemical and biological properties together with climate, inputs of labour, materials and management practices are involved in it. Management of soil conditions by man will help to modify some of these productivity factors although many will be outside man's capacity to change them.

Productivity factors may be broadly grouped into : (i) inherent factors, and (ii) factors which are dependent upon soil condition.

Inherent Productivity Factors

Plant roots require a good and firm anchorage, an adequate and balanced supply of water, air and inorganic nutrients and that there should be no impediment to their growth.

Soil texture as already stated influences all aspects of root development. A single-grain sandy texture will not easily provide a firm anchorage and especially in the seedling stage of the plant, the surface may be moved by winds, causing serious damage. On the other hand, a compact heavy soil may offer physical impediment to root-growth and thus reduce growth potential. Texture also influences the moisture status of the soil and soil-air composition and has a bearing on soil temperature.

The effective depth of the soil profile as stated earlier is another inherent feature which influences soil productivity, deeper soil enlarges the territory through which the roots can spread and increases the potential water and plant nutrient supplies.

The susceptibility of shallow soils to drought in dry periods may limit growth of many kinds of crop plant.

Soil productivity may also be affected by the presence or absence of a water table and its position. The natural drainage properties of the soil are associated with texture and water table. Poor drainage leading to anaerobic environment in which roots may suffer from high carbon-dioxide and low oxygen levels, or they may be injured by microbiologically produced compounds such as ferrous compounds or excessive quantities of soluble manganese which are harmful to them.

Parent (inorganic) material influences the natural supplies of inorganic nutrient elements. The organic matter level is of great importance on a variety of counts as already stated. Humus with its high moisture-absorbing capacity and its base/exchange properties helps to maintain adequate growth conditions, its buffering ability to oppose rapid changes in the pH of the growth medium helps to preserve evenness of soil reaction and its composition. As a direct source of organic matter, it enables the development and maintenance of a large microbiological population which is responsible for the plant's nutrient nitrogen.

Different Soil Conditions Influencing Productivity

The soil acidity as stated before is most important. Although within the range of crop plants some can be found to tolerate any particular reaction, the majority of economically valuable plants prefer a near to neutral soil. Soil reaction is usually within the control of the farmer.

The soil structure as discussed earlier affects temperature, moisture status and the soil aeration. Soil textures may be improved by appropriate measures. The incorporation of organic matter in open sandy soils will give better texture and structure, whereas the incorporation of organic matter with impervious clays will help to open them up and improve their drainage status.

By the judicious use of fertilisers and manures, the deficiencies of nutrients can often be made good. The wise management with respect to choice of appropriate crop and rotation, with the use of suitable cultural methods, can frequently minimise inherent features of a soil which would normally be considered disadvantageous. A brief review of some principal operations employed to maintain and enhance the crop-producing power of soils is given below :

Drainage : A soil containing excessive water automatically involves inadequate aeration, lowering of temperature and limited root range. All these conditions are directly related to seed germination and crop growth, and the insufficiency of air and the reduced temperature militate against the decomposition of organic matter.

Ploughing : Ploughing is the process in which the upper layer of soil is turned over. The effect of this is to bury surface organic matter and incorporate its decomposition products with the soil. The zig-zag surface produced by ploughing reduces the losses due to run-off and evaporation. Moreover the exposure of the furrows increases the formation of soil crumbs and aids the breaking down of the clods of a heavy soil with a consequent improvement of textural condition.

Rolling : The particles and their aggregates are pushed together by rolling the soil and the conditions become favourable for the water retention on the surface.

Bare Fallowing : Leaving land uncropped for a season is a very long-established practice. An enhanced production of easily available nitrogen compounds in uncropped land is one important factor for the bare fallowing.

Addition of Organic Matter : The consequent addition of humus to the soil has important effects on the soil texture, structure, micro-organic population and fertility of the soil already stated. Organic matter is added to the soil in various ways — by green manuring, by the application of farmyard manure and so forth.

Liming : On heavy soils the action of lime is to aggregate the particles and improve the structure. Increased lime status improves the conditions of microbiological development and stimulates bacterial activity. Large applications of quick lime sometimes affect a partial sterilisation of the soil.

Addition of Fertilisers : The elements required by growing plants are fairly numerous, but fertilisers containing compounds of nitrogen, phosphorus and potassium are those most commonly used in practice and those from which increased yields are most likely to arise. The use of these fertilisers is largely based on empirical experience and it is one of the pressing industrial problems which besets soil chemists in the determination of what fertilisers are likely to be most useful, and in what quantities they should be applied to any given soil area.

Relation between Soil Fertility and Soil Productivity

The term "soil fertility" should not be confused with the term "soil productivity", though both of them are sometimes used in a synonymous sense. A fertile soil is one which is rich in available plant nutrients to be taken by the plants, throughout their growth period. Thus it depends on those factors in the soil that determine its crop production potential. A productive soil, on the other hand is one which has the capacity to produce good crops. This capacity to produce good crops is derived not only from the fertility of the soil, but also from other ancillary factors such as availability of moisture, facilities for drainage, soil depth, etc.

A soil may be fertile and yet it may not be productive. A waterlogged soil may be highly fertile but may not produce good crops because of the unfavourable physical and other moisture conditions. Similarly, a fertile soil may have saline, alkaline or boron salts which being toxic to plant growth restrict its potential power to produce crops. Soils which are inherently of low fertility can be made productive through adequate management practices. The application of fertiliser and manures can even make low-fertility soils highly productive soils which are inherently of high fertility but not adequately productive because of adverse environmental conditions. Such soils can also be made productive through proper management measures. This is being done by making provision of adequate water supplies through irrigation. Other factors which restrict crop production like drainage impedance, restriction of plant-root penetration, toxic soil conditions, etc., can be overcome by adoption of suitable management practices or amelioration measures.

To sum up, soil fertility is the potential status of a soil to produce crops; while soil productivity is the resultant of various factors influencing soil management. Thus soil productivity is a function of environmental factors combined with soil fertility or more correctly it can be said that soil fertility in combination with environmental factors and management practices constitutes soil productivity.

Conclusion

In the countries, particularly in the developing ones, which have a high pressure of population on the land and find no alternative to meet the demand for food and other agricultural raw materials except through increase in agricultural production per unit land per unit time, the importance of soil fertility as well as soil productivity is being increasingly recognised.

The productivity of soil is dependent on the soil fertility. Normally, fertile soil gives good productivity. Every year soil is losing natural fertiliser through the process of production. So it is necessary to study the fertility and productivity of soil carefully to use soil rationally as well as intensively, thereby the agricultural production would be able to keep pace with the increasing population.

REFERENCES

Agarwal, R. R., 1965 : *Soil Fertility in India*, Asia Publishing House.

Ahn, Peter M., 1974 : *West African Soils*, Oxford University Press, London.

Bear, Firman E., 1965 : *Soils in Relation to Crop Growth*, Reinhold Publishing Corporation, Chapman and Hall, Ltd., London.

Beck, K. J. and R. Costa De Lemos, 1965 : *Suitability of Soil and Environment for Agricultural Production*, F.A.O., Rome.

Best, R., 1962 : "Production factors in the Tropics", *Neth. J. Agric. Sci.*, 10.

Comber, M., Willcox, J. S. and Townsend, W. N., 1960 : *An Introduction to Agricultural Chemistry*, Edward Arnold (Publishers) Ltd., London.

Dakshinamurti, C., 1965 : *Int. Atomic Energy Agency Tech. Report Series No. 48.*

De, N. K., 1978 : "Soil Fertility concepts, Problems and Solution", *Landscapes*, Deptt. of Geography, Burdwan University.

De, N. K. and Bandyopadhyay D., 1978 : "Soil Fertility — Midnapur Dist.", *Landscapes*, Dept., of Geography, Burdwan University.

Dhar, N. R., 1972 : *World Food Crisis and Land Fertility Improvement*, University of Calcutta.

Finck, A., 1973 : "The fertility of tropical soils under the influence of agricultural landuse", *Applied Science and Development*, Vol. I, The Inst. for Scientific Co-operation.

Govinda Rajan, S. V. and Gopala Rao, H. C., 1971 : *Soil and Crop Productivity*, Asia Publishing House.

Joffe, Jacob S., 1965 : *The ABC of Soils*, Oxford Book Company, Calcutta.

Kanwar, J. S., 1976 : *Soil Fertility : Theory and Practice*, ICAR, Delhi.

Keen, B. A., 1931 : *The Physical Properties of Soils*, Longmans Green and Col. Ltd., London.

Raychaudhuri, S. P., 1966 : *Land and Soil*, National Book Trust, India, New Delhi.

Riquier, J. P., Cornet J. P. and Bramao, D. L., 1965 : *A New System of Soil Appraisal in Terms of Actual and Potential Productivity*, F.A.O., Rome.

Shaw, B. T., 1952 : *Soil Physical Conditions and Plant Growth*, Academic Press Inc. Publishers, New York.

Soil Survey Staff, 1966 : *Soil Survey Manual*, U.S. Dept, Agricultural Handbook No. 18. Oxford and IBH Publishing Co., Calcutta.

Stories, R. E., 1933 : "An index for rating the agricultural value of soils", *Calif. Agric. Exp. Sta. Bull.*

Tamhane, R. V., Motiramani, D. P., Bali, Y. P. and Donahue, R. L., 1964 : *Soils, Their Chemistry and Fertility in Tropical Asia*, Prentice-Hall, New Delhi.

Thompson, Louis M., 1957 : *Soils and Soil Fertility*, McGraw Hill Book Company.

Thompson, L. M., and Troeh, F. R., 1975 : *Soils and Soil Fertility*, New Delhi.

Townsend, W. N., 1973 : *An Introduction to the Scientific Study of the Soil*, London.

Watson, K. H. and Goldsworthy, P. R., 1964 : *Soil Fertility Investigations in the Middle Belt of Nigeria*, Eup. J. Exp. Agric. 32.

[illegible] R. [illegible] (19[illegible]) *The Physical Properties of Soils*. [illegible] and Co., London.

Raychaudhuri, S. P. [illegible] *Land and Soils*. National Book Trust, India, New Delhi.

[illegible] S. E. [illegible] *Introductory Soil [illegible]* [illegible]

Stevenson, F. J. [illegible] *Soil Science and [illegible]*. [illegible] Publishers, New York. [illegible]

[illegible] Survey Staff [illegible] U.S. Dept. Agriculture [illegible] Handbook No. [illegible]

Oxford and IBH Publishing Co., Calcutta.

Singh, R. [illegible] *[illegible] saline and alkali soils*. [illegible]

[illegible] pp. [illegible]

Tisdale, S. L., Nelson, W. L., Beaton, J. D. and [illegible] *Soil Fertility and Fertilizers*. [illegible] Prentice Hall [illegible]

Thompson, L. M. [illegible] *Soils and Soil Fertility*. McGraw-Hill Book Company, [illegible] New York.

[illegible]

[illegible] S. [illegible] *Soil [illegible]*

Waksman, S. A. [illegible]

18

ECOLOGICAL APPROACH TO LAND CAPABILITY CLASSIFICATION AND REGIONAL PLANNING

R. K. GUPTA

LAND CAPABILITY is generally referred to as an expression of all the environmental parameters as they apply to the biological potential of a given resource from a specific unit of land. With finite resource capabilities and potentially infinite demandwise landuse, planning must be constrained by the inherent capability of the land to produce resources. Although this concept is currently well accepted, the approach to assess land capability is quite variable. Odum (1971) states that

> "Landuse is everybody's business and the application of ecological principles to landuse planning is undoubtedly the most important application of environmental science."

The concept widely used in the country (Tejwani, 1976) takes into consideration primarily the parameters of soil such as texture of the top soil, effective soil depth, associated land features, such as the slope of land, extent of erosion, degree of wetness and susceptibility to overflow and flooding. The capability class is determined by the degree of limitation in landuse together with the hazards involved. Within a land capability class, sub-classes are determined by the kind of limitation on hazard such as susceptibility to erosion, excess of water or soil limitations within the rooting zone or the climatic limitations like temperature or lack of moisture. Because of the peculiar socio-economic constraints, such as the size of holdings, dependence of a large population on agriculture, these concepts for landuse planning cannot strictly be applied in practice. Therefore, an ecological approach to land capability classification needs to be evolved and discussed for more pragmatic classification of land. Some of these aspects have been reviewed in this paper. India, being a vast country, has enormously varying physiographic, soil, climatic and socio-economic conditions and situations. Thus, no uniform system of classification is feasible which is acceptable to users. This paper records some observations primarily for the Western Himalayas.

Terminology and Concepts

Foresters commonly use the term "site" to report physical location and environmental features of a forest ecosystem (Spur and Barnes, 1973). The terms "habitat" and "site" are often considered as approximate equivalents (Ford Robertson, 1971). The term "site" has been defined as "area" considered in terms of its environment, particularly as this determines the type and quality of the vegetation the area can carry. Site types are classified qualitatively by their climate, soil and vegetation according to their source. This is a broader definition than the Finnish system of site types (Cajander, 1976) based on used undergrowth species, also as environmental indicators. "Habitat type", as a site classification tool, has been considered of considerable value since potential climax vegetation per se is of little interest. Although sites within a habitat type have essentially biotic potential (Daubenmire, 1968), this cannot be extended to say that all sites within a habitat type have equivalent capability to produce all resources. The habitat type approach is not structured to directly predict specific resource capabilities or specific environmental factors (Pfister, 1975).

Integrated land surveys initiated in Australia devised a "land system" which has been defined as an area or group of areas throughout which there is a recurring pattern of topography, soils and vegetation (Christian & Stewart, 1953). Integrated mapping based on the concept of biogeography (Gaussen, 1957; Rey, 1961; Emberger, 1959; Emberger and Long, 1959) and ecosystem (Sukashev, 1944; Tansley, 1955; Odum, 1971) are being followed by different countries throughout the world and reviewed by Gupta (1972). In both these methods, importance is given to vegetation. The land form, soil and aspect of biotic environment are emphasised as inset maps while composite mapping is seldom done. Recently the concept of Unique Mapping Area (UMA) has been developed in which boundaries of soil, land form, vegetation, geology, landuse and water resource is delineated having recurring patterns of land form, soil and vegetation. The other concept is MLRU which has been defined as an area or group of areas having : (a) recurring pattern of soil, land form and vegetation; (b) recurring pattern of socio-economic condition; and (c) recurring pattern of human activities.

The concept is thus based on the homogeneity of the conditions that affect human life in a particular area. These concepts are similar to the concept of "ecological response unit" or "biological potential unit". This unit has also been termed as agro-ecological zone (Gupta and Saxena, 1971).

The floral composition of a site may be looked upon as an expression of site potential. The plants on a site can be an expression of soil conditions, climatic conditions such as precipitation and temperature, an expression of soil salts and even past soil disturbances. The vegetation composition can be an index to rehabilitation practices for either partial or complete rehabilitation (Gupta, 1978).

The vegetative composition on land ecosystems are currently being used as a tax base for all non-agricultural lands. The plant cover along with terrain and soil features are efficiently used to establish a land capability class from which the property tax is calculated.

Proper land management requires an understanding how specific ecosystems actually function over a period of time an understanding that is not easily obtained to be sure. The concept of maximising yield or productivity, whether a crop, water, etc., does not appear to have been all that important during a long history of natural ecosystems. Although aesthetic considerations are of minimal importance to starving people, ecological balance is of primary importance for long-term survival. No scheme to increase productivity of foodstuffs for human consumption is likely to succeed unless it takes into account the ecology of the region. Developmental schemes, such as water development in arid and semi-arid regions, though change productivity rates markedly influence the population. Structure of a whole series of native consumer species, as such that some cannot tolerate the changes and become locally extinct, while others, such as certain rodents, granivorous birds, insects and aquatic forms, can tolerate one thrive. Some of the arthropods are vectors of human diseases. The destructive effect of crop-eating insects is known and can be counteracted by various insecticides. Irrigation designers must take these into consideration and eliminate these species. If not, much, if not all, of the increased productivity may well support more rodents and birds and not more humans.

The maximisation of yield has become an economic constraint imposed on ecosystems by man, with an excess number of people already on the earth. It now appears as though the maximisation of yield is a biological necessity. From an ecological perspective, it is acceptable to maximise the yield if certain guidelines are accepted as constraints. Kassas (1978) records that "intensification of agriculture in the arid zones is certainly not desirable and, therefore, we must brain-wash ourselves that the concepts of green revolution are not applicable in the desert". Considering the needs of population, 300 million tonnes of cereals from no more than 100 million hectares — an average of three tonnes/hectares/annum in our conditions of abundant sunshine and temperature — should be possible, if cultivation is confined to only better soils. And 300 million tonnes should be more than enough to adequately feed even a population of 1,000 million, particularly when cereals will be generally supplemented by animal proteins and fruits to be produced by non-agricultural lands. As more is learned about ecosystem dynamics in specific areas, the following guidelines can be quantified for those areas :

1. Man must operate within limits of biological and ecological constraints.
2. Man-dominated systems with high population density and heavy industrialisation can be managed best by mimicking characteristics of ecosystems where man's influence is minimal.
3. Manage to maintain a relatively high level of community diversity (for all kinds of organisms). If this seems impossible e.g. where mechanised agriculture is practised, then manage at least for landscape diversity.
4. Manage to minimise nutrient loss.
5. Maintain population sizes appropriate for the habitat.
6. Recognise the management implications of "pulse phenomenon" that occur naturally in the ecosystem when relatively fee from human influences, e.g. periodic flooding, fire and pest epidemic.
7. Monitoring progress of ecosystem variables should be an integral part of land management. It provides feedback for making management decisions.

It has been felt that good landuse planning has come only after man has first severely damaged a landscape as Lepoed (1949) records, "man does not seem to be able to understand a system he did not build and, therefore, he seemingly must partially destroy and re-build before use limitations are understood". It has now been abundantly realised that success of our development programmes would be due to emphasis on a watershed* (Gupta, 1977) and the classification for use is based on natural ecological features such as soil, slope and natural biotic communities.

The earliest land reforms through settlement surveys are reported during Akbar's reign by Todarmal. Since then, these were refined during British period with the help of Cadastral mapping conducted by Survey of India. During the last five decades, land classification in the country has been carried out for purposes of land revenue settlement, land surveys for soil studies and natural resource inventories. In 1953, All Soil Survey and Soil Conservation Board undertook studies on the soils in catchment areas of different River Valley Projects (RVPs) from the aspect of land conservation. The experience gained resulted in compilation of a Soil Survey Manual which has since been revised.

Inter-State Soil Conservation Board for Punjab and Himachal Pradesh, set up in 1961, sub-divided the Beas-Sutlej catchment into self-contained sub-catchments and prepared detailed working plans for these sub-catchments. Primarily the aim of these plans is to reduce the sediments (Gupta, 1973). The working circles constituted were : (1) Arable land working circle, (2) Grassland W.C., (3) Woodland W.C., (4) Roadside stabilisation, Stream stabilisation, livestock development W.C., and (5) Silt detection W.C. Erosion intensity is determined based on the following factors :

(a) Slope;
(b) Soil depth;
(c) Soil type;

* The term watershed is synonymous with catchment or drainage basin and watershed management includes rational utilisation of land and water resources of a watershed for optimum production with minimum hazard to natural resources. It essentially relates to the soil conservation in watershed and means proper landuse, protecting land against all forms of deterioration, building and maintaining soil fertility, conserving water for farm use, proper management of local water, drainage, flood protection and sediment reduction and increasing productivity from all landuse.

(d) Density of vegetation;
(e) Presence of gullies; and
(f) Arrangement for disposal of excess water.

Eight land types are assigned carefully designated uses that can be sustained without loss of productivity. Land capability ratings are given in Table 18.1. Capability Classes I & II comprise level areas with good agricultural soils that can be continuously cultivated with precautions such as crop rotations and strip cropping. Class I land does not require any special soil and water conservation measures and has few limitations that restrict its use. Good farming practices, such as manures and fertilisers, crop rotations, etc., maintain high productivity levels. Class II lands have moderate limitations that reduce the choice of crops. It needs simple soil and water conservation practices, such as contour cultivation, contour strip cropping, graded bunds with suitable water disposal system and good farming practices like crop rotation, stubble mulching, etc. Classes III and IV lands (Steep slopes) require greater restrictions, if cultivated, such as periodic fallowing, perennial crops, etc. Due to severe limitations for use, Class III lands need intense soil and water conservation treatment coupled with careful soil management. The soil conservation practices are costly to apply and require continued attention for management and maintenance. On a moderate slope, these need contour or graded terraces or bench terraces with a water disposal system while on level, permeable, wet or water-logged soils provision for drainage is necessary. Application of bulky organic manures or green manures, etc., is necessary to improve soil permeability. In irrigated areas, use of these lands is limited by high water table, low permeability and moderate salinity or alkalinity. For Class IV lands, limitations for use are severe, soil and water conservation practices are more difficult to apply and maintain. In semi-arid conditions, though the soils may be deep and land level productivity is limited due to uncertainty of rainfall and wind erosion hazard, while in humid and sub-humid regions water erosion hazards are severe due to steep slopes and shallow soil. In low-lying areas the soil, though level, is wet and water-logged. It takes a long time to drain and be ready for cultivation, choice of crops is severely limited. Selection of soil and water conservation practices for lands I–IV are described by Tejwani (1976).

TABLE 18.1: Land capability class rating

Class	Texture	Soil depth (cm)	Alluvial soils	Black soils	Red soils	Deep Red soils of Eastern Ghats, Western Ghats and Nilgiris	Himalayas	Effect of past erosion	Susceptibility to erosion e.g., distance from active gully heads	(mm/hr)	Conductivity (mmhos/cm at 25°, 12°)	Climate
1	2	3	4	5	6	7	8	9	10	11	12	13
I.	siel; cl; 1; sl; sil; scl.	> 90 (d_5)	0-1(A)	0-1(A)	0-1(A)	0-1(A)	0-1(A)	Up to 1/4 of top soil lost; sheet erosion (e_1)	Very far away (> 60 m)	Moderate (20-62.5)	0-2	Humid climate with well-distributed rainfall throughout the year.
II.	siel; ch; sl; scl; sil,	45-90 (d_4)	1-3(B)	1-3(B)	1-3(B) 3-5(C)	1-3(B) 3-5(C)	1-3(B)	Up to 1/4 of top soil lost; sheet erosion	Minimum 60 m	Mod. slow (5-20) Mod. rapid (62.5-125)	2-4	Humid climate with occasional dry spells; sub-humid; crop yield frequently reduced by drought.

(Contd.)

TABLE 18.1 *(Contd.)*

1	2	3	4	5	6	7	8	9	10	11	12	13
III.	sc; sic; o; ls.	22.5-45 (d_3)	3-5(C) 5-10(D)	3-5(C)	5-10(D)	5-10(D) 10-15(D)	3-5(C) 5-10(D)	Up to 1/4 to 3/4 top soil lost; rill erosion (e_2)	Between 6-60 m 0-3% slope	Slow (1.25-5) rapid (125-250)	4-8	Sub-humid, crop yield frequently reduced by drought.
IV.	c; s.	7.5-22.5 (d_2)	10-15(E)	5-10(D)	10-15 (E) 15-25 (F)	15-25(F) 25-33(G)	10-15(E) 15-25(F)	3/4 top soil and up to 1/4 sub-soil lost small gullies (e_3)	—	Very slow (< 1.25) Very rapid (> 250)	8-16	Semi-arid and arid.
V.	Same characteristics as Class I land except for one or more limitations of wetness or stoniness or rockiness or adverse climatic conditions. It has no hazard of erosion like Class I land.											
VI.		7.5 or less (d_1)	15-25 (F)	10-15 (E)	25-33 (G)	33-50 (H)	25-33 (G)	Gullied land or sand dunes (e_4)	Marginal land (6m wide strip near gully head)	—	> 16	
VII.		7.5 or less (d_1)	25-33 (G)	15-25 (F)	33-50 (H)	50-100 (I)	33-50 (H) 50-100(I)	-do-	Gully-sides and beds	—	—	
VIII.		Rock	> 33 (H)	> 25 (G)	> 50 (I)	> 100	> 100	Bad lands	Gully-sides and beds	—	—	

(After Tejwani, 1976)

Classes V–VII lands are considered not suitable for cultivation and hence are recommended for use as permanent pastures, tree crops are retained in their natural state (for naturally developed forests and wildlife, etc.). Type VIII (steep slopes, thin soils) is productive only in its natural state, as habitat for game, forest produce or is valued for recreational, scenic watershed production, or other uses such as wildlife habitat and catchment for water harvesting. Marshes and swamps are classified as Class VIII, because their value as water reservoirs and wildlife habitat outweighs their value as reclaimed farmlands.

Though Class V lands have all the characteristics of Class I land, limitations of wetness or stoniness/rockiness or adverse climatic conditions make it unsuitable for crops. There are no limitations of its use for grazing, forestry and pasture development. These lands, like Class I lands, are not subjected to more than slight wind or water erosion and do not require any special soil and water conservation practices except good management, such as seeding, fertilisation and controlling water through contour furrows, drainage ditches, diversions/waterspread, etc. Low-lying areas subjected to frequent overflows, level or nearly level, stony or rocky deep soil, are examples of such lands. Class VI land has the same limitations as Class IV land except that it is more severe, is more steep, more eroded (often gullied), soils are very shallow, either too wet or too dry. It has moderate limitations for grazing and forestry. Class VII land has severe limitations for forestry and grazing since it is very steep, very severely eroded, cut up into gullies and is either too wet or too dry. The land is best utilised under forest and permanent vegetation and for limited grazing. For good management practices such as grazing within safe limits according to carrying capacity, suitable grazing management practices, fencing and watering for livestock need be applied, gullies be controlled. In forest lands, practices such as fire prevention, careful felling cycle and methods of logging need be followed.

Examples of Class VIII lands are : Very steep, rough, barren lands, rock outcrops, sandy beaches, deserts, mine tailings, etc., which do not give economical returns. Therefore, such lands are recommended to be kept under permanent cover. Soil and water conservation practices suited to Class V to VII lands are described by Tejwani (1976).

In the hilly regions, the land settlement report classify land as *Upraon* (unirrigated upland) and *Talaon* (Irrigated lowlands). The upland is further classified as *Abbal* (Class I), *Doyam* (Class II) and *Soyam* (Class III). Lands with intermittent cultivation are called *Khil* or *Katil*. The *Talaon* or lowland permanently irrigated from river is called *seras* while lowlands occasionally irrigated are called *pachars*. While concepts of land capability classification were rigidly applied for watershed management practices in a Himalayan catchment (Anonymous, 1976, 1977), the entire Himalayan region would conform to Classes V, VI, VII and VIII lands. Om Prakash (1976) has recorded (Table 18.2) that in the two Naurar catchments, only 64.7 and 44.3 per cent of the total area is fit for cultivation with moderate to intensive management care and conservation practices. Classes I and II lands are not found at all in the area. Rest of the land, i.e., Classes V-VIII, must be spared for permanent vegetation like orchards, forests pastures, etc.

TABLE 18.2: Area under different capability classes in Naurar Catchment — Kumaon Hill

Land capability classes	Catchments I		II		Conservation Needs
	Area (ha)	% of total land	Area (ha)	% of total land	
III	16.7	14.8	4.3	9.7	Moisture conservation through mulches, addition of organic manures, compost, terraces maintained.
IV	23.0	20.4	8.1	18.3	-do-
IV	33.3	29.5	7.2	16.3	Practices save plus reduction of slope.
Total		64.7		44.3	
VI	15.3	13.5	21.1	47.6	Moisture conservation and erosion control through permanent vegetative controlled grazing.
VII	18.3	16.8	1.2	2.7	Good range and forestry management practices, management of grazing, logging exclosure for promotion of vegetation.
VIII	3.3	2.9	2.4	5.4	Maintenance of maximum vegetation cover for erosion control and exclusion of grazing.
Others	3.75	3.29	0.07	0.14	Intensive care through engg. structures like levelling, retaining walls, side walls and permanent vegetation.
	113.0	100	44.5	100	

After Om Prakash, 1976.

Though cultivation is not permissible on slopes of more than 30 per cent there is an extension of area under crops obviously to feed the growing population. Always there is hunger of land for cultivation which is associated with high population density, fragmented and small holdings and low productivity. Slopes of 30–100 per cent are sometimes put under crops and are well irrigated, levelled and protected with stone risers. The physical and topographic features associated with economic considerations for bench-terracing (costing Rs.12,000 per hectare), steep slopes vis-a-vis production potential, put severe limitations with the result that most of the fields are not properly terraced. Fields where irrigation is available are better maintained and managed than unirrigated terraces.

It would be of interest to note that agriculture in the hills of Garhwal developed only during the last 100 years. In Chamoli district, between 1842 and 1914, area under agriculture increased from 103,378 acres to 283,156 acres, in Kumaon between 1842 and 1969, it increased from 172,912 acres to 300,026 acres. In the former Tehri State, agricultural land did not increase much because of high values of *Najrana.* Between 1924 and 1964, in Uttarkashi district in Rajgarh tehsil, area under agriculture increased to about 51 per cent while in Handa tehsil, about 65 per cent. In Tehri, the increase had been about 58 per cent. The increase in area of agricultural land is reflected on the area under forest cover (Sundarlal Bahuguna : personal communication). While applying the concepts of land capability in hill regions, the following problems have been faced :

1. Presently, cultivation is carried out on all slope classes. What percentage of slope, soil depth and texture should be limited for allowing cultivation in the hills keeping in view the socio-economic conditions?
2. Though a major part of the area comes under Class VIII land, what criteria should be fixed to delimit these lands in the hills?
3. Should stoniness be kept as a criteria with more weightage?
4. If irrigated areas, what should be the criteria? Slope percentage. Should it be as above or different?
5. Though some of the forests are designated as "protected forests", what is the criteria on which these are classified?
6. Should there be a limit of slope above which no logging and felling should be permitted? If so, what should be the limit. Should such areas forming the catchment of important rivers be

delineated as ecologically sensitive and not disturbed at all?

Change in landuse according to capability involves many practical difficulties. While executing the land utilisation according to capability, one of the difficulties came with Class IV lands which are at present being utilised for marginal production of cereals. A mixed landuse for orchards and cereals was found as most suitable. Soil working for crops grown as inter-crops helps in better growth of fruit plants. Another difficulty is experienced with Class VIII lands (having more than 100 per cent slopes with thin soil). As recommended, those need be fenced and protected. Experience has shown that farmers need persuasion to keep them as reserve for at least a few years since exclosures put hard pressure on their grazing rights, the number of animals being higher than the carrying capacity. Classes VI and VII lands are generally under reserve forests which normally are managed according to working plans. However, they also show some patches with poor canopy cover. Improvement of vegetative cover on Civil and Soyam lands because of rights of adjoining villagers, however, pose some practical problems. Classes IV and VI lands are generally recommended for orchards. With cultivation as a regular landuse the farmers agree for keeping their lands under orchard. Whenever irrigation could be provided these lands need be put under crops. Cash crops such as ginger, turmeric, colocasia, chillies, tomatoes, have a wide scope and improve cost creditability of the people. Spacing of fruit plants needs adjustment for optimum production. Classes III and IV lands, where irrigation is available, are being managed by farmers for optimum production depending on the capacity to invest on inputs required for purchase of HYV seeds and fertilisers. Present and proposed landuses, according to capability in a watershed at Fakot (Tehri Garhwal district) are given in Table 18.3.

Khybri (1978) suggested the depth of soil, texture of soil, soil-stone ratio, slope of land and erosion for land capability classification in the Himalayas. Since the depth of soil and slope are the most important features of the hill region, a proper limit for their use should be fixed for regional planning. The existing terraces are made on slopes steeper than 33 per cent and considering the requirements of food, it would be futile to fix this deadline. Where conditions such as soil depth and texture of soil are favourable, this limit has to be raised. However, width of terrace and suitable height of the riser which can be maintained are other considerations to be kept in mind. A terrace width

TABLE 18.3: Present and proposed landuse, land capability and treatments proposed

Present landuse	Land capability	Proposed landuse	Treatments proposed
Wastelands :	VIII	Protected area for slopes above 100% precipitous rock.	No treatment (with protection area is liable to improve).
	VII	Permanent vegetation cover under fuel and fodder reserves (slopes 50-100%).	Protection through fencing, soil conservation measures and improvement of vegetative cover in Civil and Soyam lands.
Cultivated area :			
(i) Cultivated at intervals	IV & VI	Orchards	1. Terrace restoration through grass. 2. Planting of orchard and topfeed species.
(ii) Double cropped	IV	Orchards/crops where irrigation could be provided.	1. Restoration and improvement of guls. 2. Water management through sprinkler. 3. Change in cropping patterns. 4. Terrace re-modelling.
(iii) Tripple cropped	III	Irrigated crops	1. Improved seeds, fertilisers, vegetables. 2. Cash Crops.
Forest :			
(i) No canopy forest (Open shrub/poor grass)	VI & VII	Community land improvement for fuel and fodder rescurces.	Enclosure and improvement of vegetation cover (grass-legume top soil)
(ii) Thin forest (Degraded forest cover)			
(iii) Moderately thin forest		Reserve forest	
Orchard	IV & VI	Orchard improvement	1. Water resource development and management of water for protection, irrigation by tanks, mulching, mini-catchments. 2. Training and pruning of fruit plants. 3. Planting improved stock and improving existing stock. 4. Inter-culture operations.

of four metres including shoulder bund is considered minimum for working with bullock power. Two-metre risers are being used and maintained. Since these specifications can be met up to 50 per cent, this slope limit can be raised to 50 per cent, increased slopes soil put restrictions on terrace width and riser height (45 cm effective soil depth including "C" horizon at hillside after terrace is made).

One has to seek for alternative landuses other than agriculture in the hills for proper soil and water conservation strategy for rural development and anti-food measures. The most suitable area for agriculture seems to be the valleys where irrigation facilities are available. The following criteria may be adopted :

1. Lands relatively close to the river/stream and consequently availability of water for irrigation occurring on 15-20 per cent slope with deep to very deep soils.
2. Lands having natural streams which could be used in irrigation of limited lands after meeting irrigation requirements occurring on steep and some very steep (25–33 and 33–50%) slopes with moderately deep soils (good and poor terraces).
3. Rainfed lands (depending on rains) on steeply sloping and some very steeply sloping lands (26–33 and 33–50%) with moderately deep soils.
4. Lands under (2) above with shallow soils.

Lands Not Suitable for Cultivation but Suitable for Plantation, Pasture or Forestry Uses

5. Very steep sloping lands (33–50%) with moderate soil depth.
6. Very steep sloping lands (33–50%) as above with shallow soil depth (7–20 cm), i.e., root zone limitation.
7. Lands with extremely steep slopes of over 50 per cent with severe erosion hazard with soil of moderate depth.
8. Lands with extremely steep slopes of over 50 per cent as above with very severe erosion hazard and shallow soil (7–20 cm) posing root limitations.

Lands Suitable Only for Use of Wildlife, Recreation or Protective Watershed

Highly rocky land with 70-90 per cent bedrock exposure with small scattered patches of very shallow soil interspersed with sheet rock occurring on very steep to extremely steep slopes (33–50 and over 50 per cent slopes).

The concept of protective forest at present followed by the Forest Department needs re-thinking. Probably, the priority catchments for protection have to be delineated where felling has to be completely banned.

Proper landuse planning and adoption of a national landuse policy would not only result in greater productivity from soil but will also help to solve many of our energy problems by increasing availability of fire wood, biogas as well as prolong the life of our hydel projects. The concept and practice of 3-D forest farming embraces three major items : (1) Trees valuable in themselves as a source of timber, as saviours of land and as factors in climatic amelioration; (2) Food products from trees, primarily new food for animals and same food directly for human beings; and (3) Animals living around and among the trees and feeding on their produce.

The old concept of afforestation merely for timber and fuel has to make room for the new concept of forest farming. The tree not only outyields the agricultural crop in total production, it provides the base for sound agriculture in the hill region through a system that forms a natural biological cycle into which man fits in perfectly; he can eat the food harvested from the trees and the flesh or produce of the forest-fed livestock or sell them. The manure of the animals is returned to the soil and encourages healthy and vigorous growth of plants, thus reducing the need for fertilisers to a minimum. This offers more than a system for satisfying the basic needs of man for food, fuel and other essentials.

This will put large masses of our rural poor on their feet on a permanent basis by engaging them in activities related to animal husbandry, horticulture, forestry, multiple cropping and generate employment by encouraging industry.

However, an administrative, legal, financial and technical infrastructure is needed for the successful implementation of this programme. It has to be backed by an informed public opinion and stern political will.

REFERENCES

Cajander, A. K., 1926 : The Theory of Forest Types", *Acta For. Fenn.* 29, p. 108.

Christian, C. S. and Stewart, G. A., 1953 : General Report on Survey of Katherine-Darwine Region, CSIRO, *Aust. Land Res. Ser, No. 1*, Melbourne, Australia.

Daubenmire, R., 1968 : *Plant Communities: A Text Book of Plant Synecology*, New York.

Emberger, L., 1958 : *Principles de la Methode de Pravail an service de la carte des Groupment Vegetaux du CNRS. Acadamie de agriculture de France.*

Emberger, L. and Long, G., 1959 : "Orientation Actuelle an Service de la Carte des Groupments Vegdaux de la cartographic phystosociologique applique", *Bull. Carte phytosoloqique Ser. B.* Tom, 4 Fas. 2, pp. 119-41.

Ford Robertson, F. C., 1971 : "Terminology of Forest Science, Technology, Practice and Products", *Soc. Am. For.* Washington, DC, p. 349.

Gaussen, H., 1957 : "Les Cartes de Vegetation Inst. Francais de Pondicherry", *Trav. Sec. Sc. et Tech. T. 1*, p. 2.

Gaussen, H., Legris, P. and Meher-Homji, V. M., et. al., 1968 : International Map of Vegetation and Environmental Conditions, Sheet Kathiawar with notice ICAR, New Delhi, *Trav. Sec. Sc. et Tehn. Hors. Ser. 9*, Sheet Satpura. Ibid. Hors. Ser 10 (1970), Sheet Rajputana, Ibid. No. 17.

Graham Edward, H., 1944 : *Natural Principles of Landuse*, Oxford University Press, New York.

Gupta, M. P., 1973 : Technique for the Preparation of Working Plans for Soil Conservation in Catchments of Multi-Purpose River Valley Projects, *Soil Cons. Digest 1* (1) : pp. 8-16.

Gupta, R. K., 1971 : "Planning Natural Resources", *Land : The True Resource*, New Delhi, pp. 6-16.

———, 1978 : "Ecological Distribution and Amplitude of Some Indigenous Species vis-a-vis Their Role in Afforestation of the Arid Regions of N.W. India", *Environ. Physiol, Ecol. Plants*, pp. 423-36.

———, 1978 : "Watershed Management and Training in Institutes and Universities for High Productivity Levels from Land and Environmental Quality", *Teaching Ecol. Environ. Sci.*, India, pp. 151-61.

Gupta, R. K. and Saxena, S. K., 1971 : "Integrated Ecological Survey for Agricultural Development in the Arid Zones of India: 1. Chohtan Community Development Block in Barmer district of Rajasthan", *Ann. of Arid Zone*, pp. 85-98.

Jain, S. P. and Sharma, R. C., 1978 : *Report on Detailed Soil Survey and Land Evaluation of Village Chausali, District Almora*, pp. 1-18. Report No. 411, ICAR, Nation Bureau of Soil Survey and Landuse Planning, ICAR, New Delhi.

Kassas, M., 1978 : *Desertification*, Lecture delivered at the International Symposium on "Arid Zone Research and Development", held at CAZRI, Jodhpur, from February 14-18.

Khybri, M. L., 1978 : *Land Capability Classification for the Himalayan Region*, National Seminar on Resources Development and Environment in the Himalayan Region, Mimeo, pp. 1-6.

Leopold, Aldo, 1949 : *The Land Ethic, In : A Sand County Alnanac*, Oxford University Press, Oxford.

Major Land Resources Units (MLRU) of Bikaner District, 1974 : *Basic Resources of Bikaner District (Rajasthan)*, pp. 10-37.

Odum, E. P., 1970 : Optimum Population and Environment, A Georgian Microcosm, *Current History*, 58, pp. 355-59, 365.

Odum, E. P., 1971 : *Fundamentals of Ecology*, Philadelphia, London, Toronto.

Om Prakash, 1976 : *Problems and Strategies Related to Soils of Bhikiasen Block, Almora District*, Uttar Pradesh, pp. 1-17.

Parmar, Y. S., 1977 : *Inaugural Address*, Seminar on People and Development of Forests in Western Himalayas, pp. 1-11.

Pfister, R. D., 1975 : Land Capability Assessment by Forest Types. "America's Renewable Resource Potential : The Turning Point". *Proc. 1975 Natl. Conv. Soc. Ann. For. 1976*, pp. 312-25.

Rey, P., 1961 : "Les Bases Biogeographiques De la Restauration Fonstiere et Pastorale dans la department de 1'Audre", *Bull. Service de la Carte phytogeographique Serie A*, T-6, Fasc. 2, pp. 113-42.

Sakashev, V. N., 1944 : International Union of Forest Research Organisation, Zunich Obschiu. Biol. 5 (*Trans. Ecology* 39, 1958).

Seminar on Regional Area Planning and Resource Development, 1976 : G.B. Pant University of Agriculture and Technology, Pantnagar, Nainital, (Mimeo).

Soil Survey Manual., All-India Soil & Landuse Survey Organisation, IARI, 1971, New Delhi, 2nd edn.

Spur, S. H. and Barnes, B. V., 1973 : *Forest Ecology*, 2nd edn. New York, p. 571.

Tansley, A. G., 1955 : *The Use and Abuse of Vegetation Concepts and Terms, Ecology*, 16.

Tejwani, K. G., 1976 : "Using and Interpreting Soil Information for Land Capability, Irrigability and Range Site Classification and for Highways", *Soil Cons. Digest*, 4(2), pp. 1-34.

Odum, E. P. 1971. *Fundamentals of Ecology*. Philadelphia, [illegible].

[illegible], 1975. [illegible] *Making a Block Management Plan*. [illegible], pp. [illegible]

[illegible] Seminar on Ecology and Development of Forests in Western Himalayas, pp. [illegible]

[illegible], R. D., 1975. Land Capability [illegible] *Planning* [illegible], pp. [illegible]

Roy, R. 1961. Les Bases Biologiques [illegible] et Classification [illegible] dans le département de l'Andre [illegible], pp. [illegible]

[illegible], V. N., 1954. International Union of Forest [illegible], *Forest Ecology*, [illegible] 1956.

[illegible] Regional Area Planning and Resource Development [illegible] University [illegible] (Integrated Manual, [illegible])

Soil Survey Manual. [illegible] Soil Survey Organisation, IARI, 1971. New Delhi, 2nd edn.

[illegible] S. H. and [illegible] 1975. *Forest and Environment* [illegible]

[illegible], 1958. *Use and Abuse of Vegetational Concepts and Terms*. *Ecology* [illegible]

[illegible], R. [illegible], 1976. Using and Interpreting Soil Information for Land Capability and Range Site Classification and for Highways. [illegible], 1974.

INDEX